T0292020

Impacts and Insights of the Gorkha Earthquake

Impacts and Insights of the Gorkha Earthquake

Edited by

Dipendra Gautam

Hugo Rodrigues

ELSEVIER

Elsevier
Radarweg 29, PO Box 211, 1000 AE Amsterdam, Netherlands
The Boulevard, Langford Lane, Kidlington, Oxford OX5 1GB, United Kingdom
50 Hampshire Street, 5th Floor, Cambridge, MA 02139, United States

British Library Cataloguing-in-Publication Data
A catalogue record for this book is available from the British Library

Library of Congress Cataloging-in-Publication Data
A catalog record for this book is available from the Library of Congress

ISBN: 978-0-12-812808-4

For Information on all Elsevier publications
visit our website at https://www.elsevier.com/books-and-journals

Working together
to grow libraries in
developing countries

www.elsevier.com • www.bookaid.org

Acquisition Editor: Marisa LaFleur
Editorial Project Manager: Tasha Frank
Project Manager: Maria Bernard
Publisher: Candice Janco
Cover Designer: MPS

Typeset by MPS Limited, Chennai, India

CONTENTS

LIST OF CONTRIBUTORS

Surya P. Acharya
National Society for Earthquake Technology-Nepal (NSET), Lalitpur, Nepal

Sujan R. Adhikari
National Society for Earthquake Technology-Nepal (NSET), Lalitpur, Nepal

André R. Barbosa
Oregon State University (OSU), Corvallis, OR, United States

Ayush Baskota
National Society for Earthquake Technology-Nepal (NSET), Lalitpur, Nepal

Gopi K. Basyal
National Society for Earthquake Technology-Nepal (NSET), Lalitpur, Nepal

Govinda R. Bhatta
National Society for Earthquake Technology-Nepal (NSET), Lalitpur, Nepal

Svetlana Brzev
IIT, Gandhinagar, Gujarat, India

Suresh Chaudhary
National Society for Earthquake Technology-Nepal (NSET), Lalitpur, Nepal

Hemchandra Chaulagain
School of Engineering, Pokhara University, Pokhara Lekhnath, Nepal

Ranjan Dhungel
National Society for Earthquake Technology-Nepal (NSET), Lalitpur,
Nepal

Amod M. Dixit
National Society for Earthquake Technology-Nepal (NSET), Lalitpur,
Nepal

Rakesh Dumaru
Faculty of Engineering of the University of Porto (FEUP), Porto,
Portugal

André Furtado
Faculty of Engineering of the University of Porto (FEUP), Porto,
Portugal

Dipendra Gautam
University of Molise, Campobasso, Italy

Ramesh Guragain
National Society for Earthquake Technology-Nepal (NSET), Lalitpur,
Nepal

Ganesh K. Jimee
National Society for Earthquake Technology-Nepal (NSET), Lalitpur,
Nepal

Uddhav Karmacharya
Understanding and Managing Extremes Graduate School, Pavia, Italy

Pramod Khatiwada
National Society for Earthquake Technology-Nepal (NSET), Lalitpur,
Nepal

Narayan Marasini
National Society for Earthquake Technology-Nepal (NSET), Lalitpur,
Nepal

Luís Martins
GEM Foundation, Pavia, Italy

Niva U. Mathema
National Society for Earthquake Technology-Nepal (NSET), Lalitpur, Nepal

Khadga S. Oli
National Society for Earthquake Technology-Nepal (NSET), Lalitpur, Nepal

Bishnu H. Pandey
British Columbia Institute of Technology, Burnaby, BC, Canada

Bhuwaneshwori Parajuli
National Society for Earthquake Technology-Nepal (NSET), Lalitpur, Nepal

Achyut Poudel
National Society for Earthquake Technology-Nepal (NSET), Lalitpur, Nepal

Suman Pradhan
National Society for Earthquake Technology-Nepal (NSET), Lalitpur, Nepal

Hugo Rodrigues
RISCO Polytechnic Institute of Leiria, Leiria, Portugal

Rajesh Rupakhety
Earthquake Engineering Research Centre, University of Iceland, Selfoss, Iceland

Surya B. Sangachhe
National Society for Earthquake Technology-Nepal (NSET), Lalitpur, Nepal

Nisha Shrestha
National Society for Earthquake Technology-Nepal (NSET), Lalitpur,
Nepal

Surya N. Shrestha
National Society for Earthquake Technology-Nepal (NSET), Lalitpur,
Nepal

Vítor Silva
GEM Foundation, Pavia, Italy

Maritess Tandingan
National Society for Earthquake Technology-Nepal (NSET), Lalitpur,
Nepal

Bijay K. Upadhyaya
National Society for Earthquake Technology-Nepal (NSET), Lalitpur,
Nepal

Humberto Varum
Faculty of Engineering of the University of Porto (FEUP), Porto,
Portugal

Kai Weise
ICOMOS Nepal, Kathmandu, Nepal

Revisiting Major Historical Earthquakes in Nepal: Overview of 1833, 1934, 1980, 1988, 2011, and 2015 Seismic Events

Hemchandra Chaulagain[1], Dipendra Gautam[2] and Hugo Rodrigues[3]

[1]School of Engineering, Pokhara University, Pokhara Lekhnath, Nepal [2]University of Molise, Campobasso, Italy [3]RISCO Polytechnic Institute of Leiria, Leiria, Portugal

1.1 INTRODUCTION

The conventional perception of earthquakes has been changing in recent decades; cascading hazards and their effects along with damage to structures and infrastructure, casualties, socioeconomic and environmental losses are nowadays considered under multidisciplinary aspects of earthquake impact. In fact, the societal impacts of earthquakes are interlinked and deserve to be dealt under a multidisciplinary approach. For instance, the famine following the 1260 earthquake in Nepal is believed to be the major cause of casualties rather than the earthquake itself (BCDP, 1994). As earthquake damage and its multifaceted effects differ from community to community, historical accounts and real-time data and interpretation are the backbone for predictive models as well preparedness initiatives. Studies regarding earthquakes in Nepal started mainly after the 1988 earthquake in eastern Nepal; however, multidisciplinary interpretations related to earthquakes cannot be found in existing literature. Moreover, earthquake loss estimation and development of predictive models in local scale are limited in Nepal. Chaulagain et al. (2016) depicted earthquake loss estimation for Kathmandu valley; and other areas in Nepal still have not been investigated, so that prediction of economic losses, casualties, injuries, and building damage is not possible. Loss estimation models developed for some regions of the world may not adequately represent the scenario of the other parts of the world, hence, real-time losses are required to validate such models for local losses. Historical earthquake records play a vital role in risk reduction, as such a database is free from biases found in predictive models. To this end, a detailed account of historical earthquakes can be used as

Impacts and Insights of the Gorkha Earthquake. DOI: http://dx.doi.org/10.1016/B978-0-12-812808-4.00001-8

a basic tool for risk reduction initiatives as well to develop countermeasures based on lessons of the past events. Detailed accounts of major earthquakes in Nepal considering economic losses, impact on fatality by gender, structural and infrastructural damage, urban and rural damage do not exist to the best of the authors' knowledge. Therefore, this chapter aims to fill the gap of multidisciplinary interpretation and comparisons between the notable earthquakes that struck Nepal Himalaya since the 19th century. It presents an overview of earthquakes greater than magnitude 6.5 from 1833, 1934, 1980, 1988, 2011, and 2015 to highlight the dynamics of earthquake damage and to present some insights in the case of future seismic events.

1.2 HISTORIC EARTHQUAKES IN NEPAL

Due to the continuous convergence of the Indian plate beneath the Eurasian plate, the entire area of Hindu-Kush-Himalaya (HKH) is hit by strong to major earthquakes frequently. As Nepal is centrally located in the HKH, hundreds of earthquakes of magnitude greater than 4 occur every year. Most of such earthquakes in Nepal Himalaya are limited to magnitude 4−6, and damage usually does not occur within this range of magnitude. It is interesting to note that only earthquakes of magnitude 6.5 or above are known to cause damage in Nepal, however, it should be realized that earthquake damage is not solely attributable to magnitude rather than energy release, period of shaking, focal depth, vulnerability of building stocks, and many other factors that directly influence damage during earthquakes. Based on records of earthquakes since 1911, the frequency of earthquakes in Nepal Himalaya with the return period of each category as depicted by BCDP (1994) along with recent records since 1994 are presented in Table 1.1.

Table 1.1 Earthquakes in Nepal Since 1911										
Category	1991–1991					1994–2016				
Earthquake magnitude (M_L)	5−6	6−7	7−7.5	7.5−8	>8	5−6	6−7	7−7.5	7.5−8	>8
Number of events	41	17	10	2	1	76	8	1	1	–
Approximate return interval (years)	2	5	8	40	81	–	–	–	–	–

Source: *Modified from Building Code Development Project (BCDP), 1994. Seismic Hazard Mapping and Risk Assessment for Nepal; UNDP/UNCHS (Habitat) Subproject: NEP/88/054/21.03 Kathmandu, Nepal: Ministry of Housing and Physical, Planning, Government of Nepal; National Seismological Center (NSC), 2016. <http://www.seismonepal.gov.np/> (last accessed 30.04.17.).*

The Building Code Development Project (BCDP, 1994) highlighted that the history of known earthquakes in Nepal dates to 1255, which claimed one third of the Kathmandu valley population including the then King Abhay Malla. The intensity of 1255 earthquake in Kathmandu valley was assigned as ~X in Modified Mercalli Scale. Accounts of damage outside the Kathmandu valley for the 1255 earthquake is not available; hence, a detailed damage scenario cannot be presented. After 1255, some records of earthquakes in the 13th, 15th, 17th, and 18th centuries were also outlined by the Building Code Development Project (BCDP, 1994). The most important seismic events in Nepal since 1255 and their impacts are outlined in Table 1.2.

Apart from the reported major events, the aftershock activities are not highlighted in most of the available descriptions of the historical earthquakes in Nepal. Rana (1935) noted that three preshocks occurred on January 14, 1934, and 28 aftershocks including some strong ones followed the main shock within six days after January 15. Description of aftershock activities after January 20, 1934, cannot be found, neither can we find any description related to damage attributed to aftershocks. Due to lack of instrumentation in Nepal, most of the earthquake magnitude, damage, and intensity of distribution are inferred from the descriptions that exist in epigraph or books. Two of the notable earthquakes from 2011 and 2015 were recorded by instruments installed in the Kathmandu valley and the aftershock activities of the 2015 Gorkha earthquake are also well documented by the National Seismological Center, Nepal (for details see http://www.seismonepal.gov.np/). The main shock of the 2015 Gorkha seismic sequence was followed by 480 aftershocks of local magnitude equal to or greater than 4; and some of the aftershocks, like those of April 25, 2015 (M_W 6.7), April 26, 2015 (M_W 6.9), and May 12, 2015 (M_W 7.3), aggravated the damage throughout the affected areas. Detailed analysis of seismicity and aftershock activities following the main shock of April 25, 2015 (M_W 7.8), can be found in Chapter 2, Seismotectonic and Engineering Seismological Aspects of the M_W 7.8 Gorkha, Nepal, Earthquake, by Rajesh Rupakhety.

1.3 REVISITING STRONG TO MAJOR EARTHQUAKES IN NEPAL

1.3.1 1833 Earthquake

On August 26, 1833, a strong earthquake struck Nepal and the northern part of India. The magnitude of the earthquake was estimated as between 7.5 and 7.9 ($7.5 < M_W < 7.9$) and the effect of the

Table 1.2 Historical Earthquakes in Nepal Between 1255 and 2015

Year	Epicenter	Magnitude	Casualties	Structural Damage and Geotechnical Aspects
1255		Intensity $MMI \sim X$	One third of the Kathmandu valley population along with the then King Abhay Malla	Severe damage in residential buildings, monumental and architectural heritage
1260		–	Many people killed by the earthquake as well as the famine followed by earthquake	Damages in residential buildings, monumental and architectural heritage
1408		–	No records available	• Severe damage in residential buildings, temples • Lateral spreading/soil liquefaction
1681		–	No records available	Damage on residential buildings
1810		–	Some casualties in Bhaktapur	Significant damage in residential buildings and monumental constructions
1823		–	No	Some residential buildings damaged
1833		$M_L \sim 7.7$	414 deaths in and around Kathmandu valley	• In total 18000 buildings damaged • Around 4000 in Kathmandu valley and Banepa
1834		–	No records available	Many residential as well as monumental constructions were damaged
1837		–	No records available	Damage only in Indian state of Bihar
1869		–	No records available	No records available
1897		–	No records available	No records available
1917		–	No records available	No records available
1934	Eastern Nepal	$M_W \sim 8.1$	• 8519 total deaths in Nepal • 4296 within Kathmandu valley	• More than 200000 residential buildings, monuments and historical constructions damaged • About 81000 buildings collapsed • Almost 55000 buildings damaged in Kathmandu valley • 12397 buildings collapsed in Kathmandu valley • Soil liquefaction/lateral spreading observed in central part of Kathmandu valley and many other locations of central and eastern plains of Nepal • Cascading effects like landslides, floods due to blockade in river course aggravated the damage
1936	Annapurna	M_L 7.0	No records available	No records available
1954	Kaski	M_L 6.4	No records available	No records available
1965	Taplejung	M_L 6.1	No records available	No records available
1966	Bajhang	M_L 6.0	24	• 6544 buildings damaged • 1300 buildings collapsed

(*Continued*)

Year	Epicenter	Magnitude	Casualties	Structural Damage and Geotechnical Aspects
1980	Chainpur	M_L 6.5	103	• 25086 buildings damaged • 12817 collapsed
1988	Udaypur	M_W 6.5	721	• 66382 buildings damaged • Several cases of liquefaction reported in eastern Nepal
2011	Sikkim-Nepal border	M_W 6.9	• 6 deaths and 30 injuries in Nepal side (damage was intense in Indian side) • 2 casualties in Kathmandu valley	• 14554 buildings damaged • 6435 buildings collapsed
2015	Barpak, Gorkha	M_W 7.8	• 8790 deaths and 22300 injuries • 8 million people displaced	• 498852 buildings collapsed • 256697 buildings partly damaged • As many as 3600 landslides and avalanches • Many cases of liquefaction and lateral spreading in Kathmandu valley • Severe damage in infrastructures and lifelines

Table 1.2 (Continued)

Source: *Modified from Building Code Development Project (BCDP), 1994. Seismic Hazard Mapping and Risk Assessment for Nepal; UNDP/UNCHS (Habitat) Subproject: NEP/88/054/21.03 Kathmandu, Nepal: Ministry of Housing and Physical, Planning, Government of Nepal; Ministry of Home Affairs (MoHA) Nepal, 2011. <http://drrportal.gov.np/> (last accessed 15.04.17.); National Planning Commission (NPC), 2015. Post-Disaster Need Assessment, vols. A and B. Government of Nepal, Kathmandu, Nepal.; Gautam, D., Chaulagain, H., 2016. Structural performance and associated lessons to be learned from world earthquakes in Nepal after 25 April 2015 (M_W 7.8) Gorkha earthquake. Eng. Fail. Anal. 68, 222–243.*

earthquake was felt within 1 million sq. km (Bilham, 1995). As reported by Bilham (1995), the main shock was preceded by two strong fore-shocks five hours earlier. Per the Building Code Development Project (1994), almost 18,000 buildings were damaged in Nepal, of which almost 4,000 were within the Kathmandu valley, including hundreds of heritage structures. In total 414 fatalities were reported in the Kathmandu valley (BCDP, 1994) and cascading effects, like rockfalls and earthquake-triggered landslides, were believed to occur after the earthquake. Bilham (1995) reported that most of the people would have chosen open spaces immediately after the strong foreshocks, which resulted in relatively low fatalities, although 30% or more buildings were believed to have collapsed. Probably, choice of open spaces was also supported by folklore, because most of the people in Nepal believe that the earthquake is a reversible phenomenon and the strong shaking should be followed by another strong shaking in the reverse direction. The areas affected in Nepal by 1833 earthquake along with tentative building damage equivalent intensity are presented in Table 1.3.

Table 1.3 Areas Affected Due to 1833 Earthquakes Including Intensity and Damage Scenario		
Location	Intensity	Number of Damaged Buildings
Bungamati	IX	80 (30% destroyed)
Banepa	IX	20
Badegaon	IX	35
Bode	IX	20
Bhaktapur	X	2000 (75% of total)
Ichangunarayan	IX	20
Chapali	IX	7
Chapagaon	IX	35
Chitlang	IX	14
Dhulikhel	IX	21
Dhunibeshi	IX	40
Gorkha	VIII	2
Gokarna	IX	8
Harisiddi	IX	20
Hadigaon	IX	20
Kathmandu durbar square	IX	400
Kirtipur	VIII	14
Kerung	IX	60 (15% destroyed)
Khokana	IX	130
Lubhu	IX	25
Nagadesh	IX	20
Nala	IX	11
Panga	IX	24
Panauti	IX	19
Patan	X	285
Pyanggaon	IX	8
Phalametar and vicinity (near Panauti)	IX	300
Pharping	IX	8
Sanagaon	IX	40
Sankhu	IX	40
Thaiba	IX	18
Thimi	X	150
Thankot	IX	23
Tokha	IX	15

Source: *Modified from Bilham, R., 1995. Location and magnitude of the 1833 Nepal earthquake and its relation to the rupture zones of contiguous Great Himalayan Earthquakes. Curr. Sci. 69 (2), 155–187.*

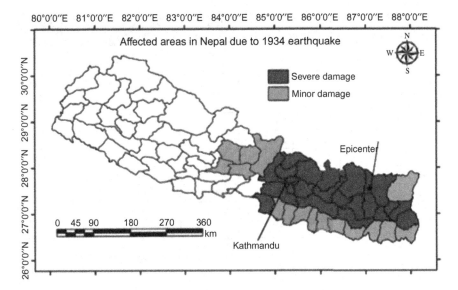

Figure 1.1 Areas affected by the 1934 earthquake.

1.3.2 1934 Bihar-Nepal Earthquake

On January 15, 1934, the greatest earthquake ($M_W \sim 8.4$) of modern times occurred in eastern mountains of Nepal (Fig. 1.1). Rana (1935) reported that the duration of the earthquake was nearly 120 seconds, whereas some other descriptions noted that the overall shaking was of 8–10 minutes. Due to lack of instrumentation and associated scientific records, exact magnitude and ground motion cannot be determined. However, per the descriptions presented by Rana (1935) and others, the earthquake was devastating especially in the eastern mountains, eastern and central plains, as well as central mountains of Nepal. A generalized damage scenario due to 1934 earthquake is presented in Fig. 1.1 for states suffering minor and severe damage.

Losses due to the Bihar-Nepal earthquake were primarily attributed to casualties and building damage, whereas some reports of rockfalls and landslides were also reported in the eastern and central mountains of Nepal. In Nepal, 8519 casualties were reported, whereas the records on Indian side depicted 7188 casualties (Rana, 1935). Altogether 207,248 cases of structural damage were reported throughout Nepal, of which 134,932 were in the eastern mountains only. In the Kathmandu valley, 19% of buildings were estimated to have collapsed and another 38% of buildings sustained considerable damage due to

Table 1.4 Summary of Casualties and Structural Damage Due to Bihar-Nepal Earthquake						
Location	Deaths		Residential Building Damage			Number of Damaged Monuments
	Male	Female	Collapsed	Major	Minor	
Kathmandu	254	225	725	3735	4146	40
Outskirt of Kathmandu	79	166	2892	4062	4267	16
Patan	250	297	1000	4170	3860	226
Outskirt of Patan	871	826	3997	9442	1598	30
Bhaktapur	433	739	2359	2263	1425	177
Outskirt of Bhaktapur	65	91	1444	1986	2388	–
East No.1	163	193	9628	19391	–	–
East No. 2	52	43	4687	10738	–	–
East No. 3	330	527	21107	15548	–	–
East No. 4	698	899	15048	5	–	–
Dhankuta	162	154	6623	15120	–	–
Ilam	41	51	2316	3112	–	–
Udaypurgadhi	295	257	1052	3917	–	–
Sindhuligadhi	51	58	3486	3154	–	–
West No. 1	4	6	582	1720	–	–
West No. 2	–	1	186	461	–	–
West No. 3	–	1	19	65	–	–
West No. 4	–	1	8	1	–	–
Chisapanigadhi	25	27	–	18	1266	–
Birgunj	16	28	3654	854	2546	–
Mahottari-Sarlahi	31	20	–	4323	268	–
Saptari-Siraha	17	23	47	428	–	–
Biratnagar	13	36	13	1	64	–

Source: *Modified from Rana, B.S.J.B. 1935. The Great Earthquake of Nepal. Jorganesh Press, Kathmandu, Nepal.*

the Bihar-Nepal earthquake (Rana, 1935). Table 1.4 presents casualties and damage to the structures in Nepal due to 1934 earthquake.

1.3.3 1980 Chainpur Earthquake

On July 29, 1980, the far western Nepal and adjoining Indian territories were struck by a strong earthquake of magnitude 6.5. As most of the notable earthquakes of modern times in Nepal Himalaya are largely concentrated in the eastern and central regions, the 1980

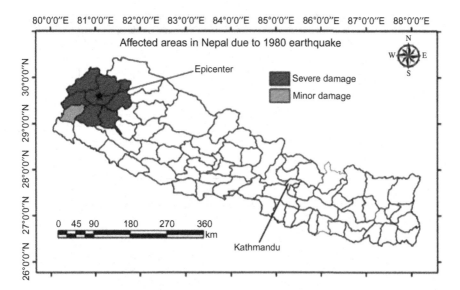

Figure 1.2 Areas affected by the Chainpur earthquake.

Chainpur earthquake is the only notable earthquake of magnitude 6.5 or above in the western section of the central seismic gap. This earthquake caused 46 casualties and 236 injuries in the affected areas of the far western mountains of Nepal. Fig. 1.2 highlights the affected areas along with the extent of earthquake damage due to the Chainpur earthquake.

Although 12,817 buildings collapsed and 13,298 sustained major damage due to the Chainpur earthquake, the fatality rate was relatively lower and that may be due to occurrence of a foreshock, as in the case of the 1833 earthquake. Damage statistics along with casualties in the affected areas is depicted in Table 1.5 as follows.

1.3.4 1988 Udaypur Earthquake

The Udaypur earthquake (M_W 6.8) struck eastern and part of central Nepal including the Kathmandu valley on August 21, 1988, at 4:54 a. m. local time (see Fig. 1.3). The earthquake caused 721 fatalities and 12,244 injuries and damaged nearly 66,000 buildings, mainly in the eastern plains, eastern mountains, and parts of the central mountains in Nepal (BCDP, 1994). The effects of the Udaypur earthquake were documented in detail by Gupta (1988), which states that the earthquake not only caused damage to many residential buildings but

Table 1.5 Summary of Damage Due to the Chainpur Earthquake

Location	Deaths	Injuries	Building Damaged		
			Collapse	Major	Minor
Darchula	24	–	4135	2743	–
Baitadi	22	236	1257	1949	–
Dadeldhura	–	–	–	120	–
Bajhang	–	–	6137	6380	2200
Bajura	–	–	419	654	1199
Achham	–	–	781	1227	1583
Doti	–	–	82	225	1395

Source: *Modified from Singh, V., 1985. Earthquake of July 1980 in Far Western Nepal. J. Nepal Geol. Soc. 2 (2), 1–11.*

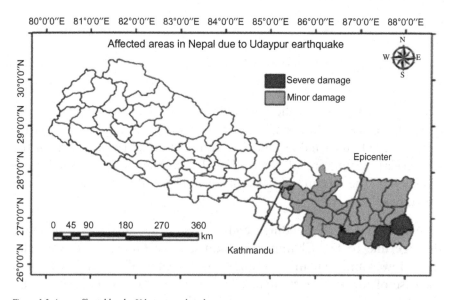

Figure 1.3 Areas affected by the Udaypur earthquake.

several other lifelines were affected, leading to huge economic losses. Altogether 1202 school buildings and 1159 public buildings were damaged by the Udaypur earthquake and hundreds of kilometers of road, irrigation canals, highway bridges, and water supply networks were also affected. Overall, the reconstruction cost for damaged buildings and lifelines was estimated to be around $66.4 million (USD) by Gupta (1988). A summary of the casualties and structural damage due to the Udaypur earthquake is presented in Table 1.6.

Location	Deaths	Injuries	Residential Building Damage		Public Building Damage	
			Collapsed	Damaged	Collapsed	Damaged
Taplejung	3	7	767	293	–	12
Panchthar	99	261	5244	6804	94	2
Ilam	73	129	5918	5538	181	88
Jhapa	–	19	31	163	4	20
Sankhuwasabha	12	46	1944	704	39	10
Bhojpur	14	88	2956	3114	21	–
Tehrathum	67	76	4481	3296	25	26
Dhankuta	93	154	7277	2384	47	59
Sunsari	138	327	2494	4466	11	22
Morang	32	941	637	852	38	38
Solukhumbu	–	4	297	341	2	1
Okhaldhunga	8	28	2162	3137	7	12
Khotang	26	140	7919	7143	10	11
Udaypur	82	70	5457	3933	27	20
Siraha	8	41	76	1279	10	58
Saptari	13	45	1263	1138	24	28
Mahottari	1	2	26	4	–	38
Dhanusha	2	3	375	306	–	55
Sindhuli	32	33	1670	1177	11	25
Dolakha	2	5	268	933	21	19
Ramechhap	2	1	589	1800	10	31
Kathmandu	–	–	–	200	–	19
Lalitpur	1	3	376	137	–	3
Bhaktapur	7	23	274	1477	–	11
Kavre	–	–	–	5	–	1
Sindhupalchowk	2	4	711	478	22	38

Table 1.6 Summary of Damage Due to Udaypur Earthquake

Source: *Modified from Gupta, S.P., 1988. Eastern Nepal Earthquake 21 August 1988, Damage and Recommendations for Repairs and Reconstruction. Asian Disaster Preparedness Center, Asian Institute of Technology, Bangkok, Thailand.; T. Fujiwara, T. Sato, H.O. Murakami, T. Kubo., 1989. Reconnaissance Report on the 21 August 1988 Earthquake in the Nepal-India Border Region, Research Report on Natural Disasters. Japanese Group for the Study of Natural Disaster Science, Tokyo, Japan.*

1.3.5 2011 Sikkim-Nepal Border Earthquake

On September 18, 2011, eastern Nepal was struck by a strong earthquake of magnitude 6.9 at 6:25 p.m. local time. The epicenter was in the area bordering Nepal and the Indian state of Sikkim and damage was particularly intense in Sikkim rather than on the Nepali side.

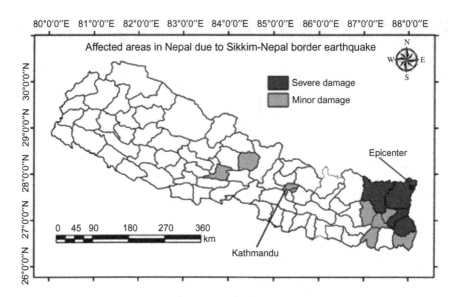

Figure 1.4 Areas affected by the 2011 Sikkim-Nepal border earthquake.

As reported by Ministry of Home Affairs, Government of Nepal (MoHA, 2011), six deaths and 30 injuries were recorded in Nepal. Out of six deaths, two were in the Kathmandu valley and the other deaths were in eastern mountains, where building collapse and severe damage was dominant. As many as 12,301 people from 4851 families were reported to be displaced due to this earthquake. In total, 6,435 buildings collapsed, 11,520 were moderately damaged, and 3,024 buildings sustained minor damage in Nepal (MoHA, 2011). The epicentral location and damage extent in the areas affected by the Sikkim-Nepal border earthquake are mapped in Fig. 1.4.

1.3.6 2015 Gorkha Earthquake

The 2015 Gorkha earthquake occurred on April 25, 2015, nearly 80 km N-NE of Kathmandu. The epicenter was in the Barpak village of the Gorkha district, and damage was concentrated primarily toward the east of the epicenter. As of April 2017, 480 aftershocks of local magnitude 4 or above were recorded by the National Seismological Center (http://www.seismonepal.gov.np/). The main shock of April 25, 2015 (M_W 7.8), was followed by a strong aftershock of magnitude 6.7 on the same day. Moreover, on April 26, 2015, another strong aftershock of magnitude 6.9 struck central Nepal, and the strongest aftershock (M_W 7.3) of the Gorkha seismic sequence was recorded on May

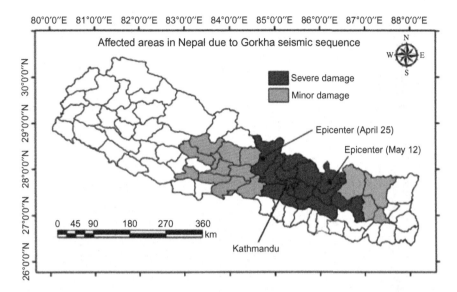

Figure 1.5 Areas affected by the Gorkha seismic sequence.

12, 2015. Damage due to the Gorkha earthquake was attributed to these four major shakings in central Nepal, which caused 8790 fatalities, 22,300 injuries, and affected 8 million people from 31 out of 75 districts in Nepal (NPC, 2015). The Gorkha earthquake caused damage to both building structures and lifelines like road networks, hydropower projects, and water supply systems. The Post-Disaster Need Assessment conducted by National Planning Commission of Nepal highlighted that the overall loss due to the Gorkha earthquake was $7 billion (USD; NPC, 2015). Fourteen out of 75 districts in Nepal were declared as the crisis-hit areas by the government immediately after the earthquake, as shown in Fig. 1.5. Seismic performance, lessons, and future insights regarding the damage to residential buildings and heritage structures are reported by Goda et al. (2015), Shakya and Kawan (2016), Gautam et al. (2016), Gautam (2017), and Gautam and Chaulagain (2016). A summary of casualties and damage is presented in Table 1.7.

1.4 EARTHQUAKE IMPACTS IN NEPAL: MULTIDISCIPLINARY PERSPECTIVES

Since early recorded history of earthquakes in Nepal, earthquake damage occurrence is understood to be anomalous both from the

Table 1.7 Summary of Casualties and Damage Due to the Gorkha Earthquake

District	Fatalities		Injuries	Number of Residential Buildings		Number of Health Facilities		Number of Damaged School Buildings	Number of Government Buildings	
	Male	Female		Collapsed	Partially Damaged	Collapsed	Partially Damaged		Collapsed	Partial Damage
Sindhupalchowk	1497	1943	1753	63885	2751	74	23	546 (98%)	710	37
Kathmandu	622	600	7952	43502	56024	11	52	187 (63%)	85	277
Lalitpur	69	108	3051	17444	8064	19	20	149 (75%)	217	198
Bhaktapur	119	214	2101	18900	9054	6	20	137 (100%)	5	51
Gorkha	215	233	952	59527	13428	40	39	495 (100%)	227	36
Sindhuli	5	10	230	18197	10028	15	44	451 (81%)	92	231
Rasuwa	312	344	771	11368	267	21	6	98 (100%)	8	4
Okhaldhunga	10	10	61	10031	3107	15	15	228 (69%)	18	38
Makwanpur	16	17	229	20035	17383	39	20	361 (68%)	46	177
Kavre	129	189	1179	49933	23714	55	76	548 (92%)	48	31
Dhading	340	393	1218	81313	3092	69	37	587 (97%)	93	58
Nuwakot	461	638	1050	75562	4200	55	44	485 (100%)	15	14
Dolakha	84	85	662	48880	3120	52	31	363 (92%)	517	–
Ramechhap	17	23	134	26743	13173	33	33	151 (32%)	54	56

Source: Modified from Ministry of Home Affairs (MoHA) Nepal, 2011. <http://drrportal.gov.npl> (last accessed 15.04.17.).

engineering and socioeconomic perspectives. It may be due to seismicity and patterns of earthquakes, too. For instance, due to the occurrence of a foreshock preceding the 1833 earthquake, fatalities should have been minimized, as the earthquake was greater than magnitude 7.5 and, as reported in Table 1.1, severe impact was noted in the Kathmandu valley, where population density was higher than any other area of Nepal. The 1934 earthquake was also preceded by three foreshocks, however, 8519 deaths occurred in Nepal, which is the most devastating event of the 20th century. The foreshocks during the 1934 earthquake should not have been felt in the Kathmandu valley and other worst affected areas, so that people were not alarmed as in the case of 1833 and 1988 earthquakes. Although average building damage percentage in Kathmandu valley was relatively equal during the 1833, 1934, and 2015 Gorkha earthquakes, the foreshock should have played a vital role in minimizing the fatality rate in case of the 1833 earthquake. In terms of gender, historical earthquakes have depicted that fatality rate is higher among females than males. For instance, during the 1934 earthquake 55% of total fatalities was among females, whereas during the 1988 earthquake this rate was 52%. During the 2015 Gorkha earthquake, the female fatality rate was greater than 55%, which depicts the unchanged scenario of women as the more vulnerable group during earthquakes since 1934. Although the magnitude of the Gorkha earthquake was at least four times smaller than the 1934 earthquake, female fatality remained higher than that of 1934 earthquake. Evidently, we can infer that in case of major earthquakes, female fatality will be very high in Nepal.

In terms of overall fatality, total fatality due to the Gorkha earthquake should be also affected due to its occurrence during midday, when most of the people were outside their home. Moreover, the main shock occurred on Saturday, so that schools, offices, and other workplaces were closed; possibly this should be one of the causes of low fatality rate during the Gorkha earthquake, too.

During the 1934 earthquake, 50% of total fatalities was reported within the Kathmandu valley, whereas the Gorkha earthquake had only 19.7% of the total fatalities in the Kathmandu valley. Although the population in the Kathmandu valley has risen more than tenfold since the 1934 earthquake, advancement in building structural forms probably contributed to the lower fatality rate. Nearly 15% of total building collapse was recorded in the Kathmandu valley during the

2015 earthquake whereas only 5% of total collapse was in the Kathmandu valley during the 1934 earthquake. Due to a lack of infrastructure, like the roadways, hydropower projects, bridges, and water supply systems, during the 1833 and 1934 earthquakes, sectoral comparisons with the 2015 Gorkha earthquake is not possible.

1.5 FINAL REMARKS

This study set out to provide the first systematic account of casualties, damage, and losses due to notable earthquakes since the 19th century in Nepal. We present the classified damage and mapped the damage extent in affected areas. Our analysis confirms that the female fatality rate during earthquakes is greater than male fatality rate in Nepal. As the 2015 earthquakes were not preceded by foreshocks, this probably may be the major cause of the high fatality rate. This study adds to the growing body of research that indicates the need for historical data in interpretation of earthquake impact. Our analysis confirms that anomalies in terms of structural damage, casualties, and economic losses played a dominant role during each Himalayan earthquake. Moreover, we conclude that the occurrence of foreshocks ultimately reduces the fatalities in Nepal.

REFERENCES

Bilham, R., 1995. Location and magnitude of the 1833 Nepal earthquake and its relation to the rupture zones of contiguous great Himalayan earthquakes. Curr. Sci. 69 (2), 155–187.

Building Code Development Project (BCDP), 1994. Seismic Hazard Mapping and Risk Assessment for Nepal; UNDP/UNCHS (Habitat) Subproject: NEP/88/054/21.03. Ministry of Housing and Physical, Planning, Government of Nepal.

Chaulagain, H., Rodrigues, H., Silva, V., Spacone, E., Varum, V., 2016. Earthquake loss estimation for the Kathmandu valley. Bull. Earthq. Eng. 14 (1), 59–88.

Fujiwara, T., Sato, T., Murakami, H.O., Kubo, T., 1989. Reconnaissance Report on the 21 August 1988 Earthquake in the Nepal-India Border Region, Research Report on Natural Disasters, Japanese Group for the Study of Natural Disaster Science, Tokyo, Japan.

Gautam, D., 2017. Seismic performance of world heritage sites in Kathmandu valley during Gorkha seismic sequence of April–May 2015. J. Perform. Construct. Facil.10.1061/(ASCE) CF.1943-5509.0001040, 06017003.

Gautam, D., Chaulagain, H., 2016. Structural performance and associated lessons to be learned from world earthquakes in Nepal after 25 April 2015 (M_W 7.8) Gorkha earthquake. Eng. Fail. Anal. 68, 222–243.

Gautam, D., Rodrigues, H., Bhetwal, K., Neupane, P., Sanada, Y., 2016. "Common structural and construction deficiencies of Nepalese buildings.". Innov. Infrastruct. Solut. 1 (1), 1. Available from: http://dx.doi.org/10.1007/s41062-016-0001-3.

Goda, K., Kiyota, T., Pokhrel, R., Chiaro, G., Katagiri, T., Sharma, K., et al., 2015. "The 2015 Gorkha Nepal earthquake: insights from earthquake damage survey." Front. Built Environ. 1, 8.

Gupta, SP., 1988. Eastern Nepal Earthquake 21 August 1988, Damage and Recommendations for Repairs and Reconstruction. Asian Disaster Preparedness Center, Asian Institute of Technology, Bangkok, Thailand.

Ministry of Home Affairs (MoHA) Nepal, 2011. <http://drrportal.gov.np/> (last accessed 15.04.17.).

National Planning Commission (NPC), 2015. Post-disaster Need Assessment, vol. A and B. Government of Nepal, Kathmandu, Nepal.

National Seismological Center (NSC), 2016. <http://www.seismonepal.gov.np/> (last accessed 30.04.17.).

Rana, B.S.J.B., 1935. The Great Earthquake of Nepal. Jorganesh Press, Kathmandu, Nepal.

Shakya, M., Kawan, C., 2016. "Reconnaissance based damage survey of buildings in Kathmandu valley: An aftermath of 7.8 M_w, 25 April 2015 Gorkha (Nepal) earthquake." Eng. Fail. Anal. 59, 161–184. Available from: http://dx.doi.org/10.1016/j.engfailanal.2015.10.003.

Singh, V., 1985. Earthquake of July 1980 in Far Western Nepal. J. Nepal Geol. Soc. 2 (2), 1–11.

CHAPTER 2

Seismotectonic and Engineering Seismological Aspects of the M_W 7.8 Gorkha, Nepal, Earthquake

Rajesh Rupakhety
Earthquake Engineering Research Centre, University of Iceland, Selfoss, Iceland

2.1 INTRODUCTION

The 2015 Gorkha earthquake occurred on April 25 at 11:26 a.m. UTC, local time 11:56 a.m. The earthquake ruptured a segment of the Main Himalayan Thrust (MHT) fault, a low-angle continental subduction interface between the Indian Plate to the south and Eurasian Plate to the north. The epicenter of the earthquake was located near Barpak village in the Gorkha district of Nepal, approximately 80 km WNW of Kathmandu (see Fig. 2.1). According to the Global Centroid Moment Tensor (GCMT; Ekström et al., 2012), the moment magnitude of the earthquake was 7.8. The National Seismological Centre (NSC) of the Department of Mines and Geology (DMG) of Government of Nepal assigned a local magnitude of 7.6 (NSC, 2017).

The earthquake and its aftershocks caused widespread destruction in the central and eastern parts of Nepal. Official records (NDRRIP, 2015) report 8856 human casualties, 22,309 injured people, 2673 fully damaged government buildings, 3757 partially damaged government buildings, 602,257 fully damaged private houses, 285,099 partially damaged private houses, 8038 affected schools, 19,708 fully damaged classrooms, and 11,046 partially damaged classrooms. The number of casualties and injuries was much less than what could have been expected for an event of this size. The main reason for this is the fortunate timing of the earthquake. It occurred at noon on a Saturday, when schools were closed and most of the people and children in the villages were outside of their dwellings. Considering 19,708 fully damaged classrooms, the number of casualties could have been much higher had the earthquake occurred on a school day. The total numbers of

Impacts and Insights of the Gorkha Earthquake. DOI: http://dx.doi.org/10.1016/B978-0-12-812808-4.00002-X

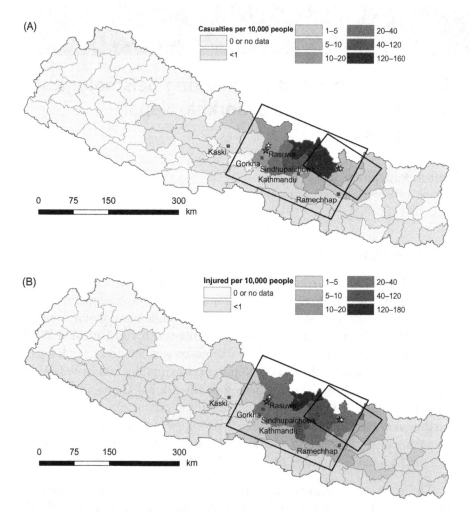

Figure 2.1 Distribution of (A) casualties and (B) injuries caused by the 2015 Gorkha Earthquake and its after-shocks. The epicenters of the main shock inferred by the United States Geological Survey (2015) and NSC are indicated by red star (gray in print versions) and white star, respectively, in the Gorkha district. The epicenters of the largest aftershock (moment magnitude 7.3) are indicated by stars in the Dolakha district (north of the Ramechhap district). The rectangles indicate the boundary of fault plane solutions of the two earthquakes as estimated by USGS (2015). Injuries and casualties are reported in number of people per 10,000 based on census data from 2011.

fully damaged houses also indicate that the casualties could have been many times higher had the earthquake occurred at night. The district-wise distribution of casualties and injuries are shown in Fig. 2.1A and B, respectively. In terms of casualties relative to the population, Rasuwa and Sindhupalchowk suffered the most. Districts north of Kathmandu and lying between the epicenters of the main shock and the largest aftershock on May 12, 2017 (moment magnitude 7.3), were the most severely affected ones. The proportion of population injured in Kathmandu, Bhaktapur, and Latitpur was relatively high. The difference in the proportion of dead and injured people between these districts and other districts to the north and east is due to the relatively better performance of buildings: total collapse of buildings in these districts was less frequent than in districts such as Rasuwa, Sindhupalchowk, Dolakha, Nuwakot, and Dhading. The most severely affected areas are within the ruptured fault planes of the two earthquakes, and more on the northern side, because most of the slip on the fault planes occurred on the lower edge of the north-dipping MHT. Unilateral propagation of rupture to the east from the epicenter of the main shock caused the districts east of Gorkha to be more severely affected that those to the west.

This chapter presents a brief overview of the seismotectonics of the Himalayan region near central Nepal in the context of the Gorkha Earthquake and its aftershocks. Seismotectonics of the region is briefly explained with reference to large earthquakes of the past. The source process of the main shock, as revealed by numerous studies based on teleseismic as well as geodetic data is reviewed in some detail. Spatial and temporal characteristics of the aftershocks are discussed. Some salient features of ground motion recorded in the Kathmandu Valley are discussed, and some recommendations for future research are provided.

2.2 SEISMOTECTONICS AND SEISMICITY

The seismotectonics of the region is controlled by the collision of the Indian and the Eurasian tectonic plates. The Himalaya was created by the uplift caused by this collision, while subsidence to the north created the Tibetan Plateau. The Himalaya extends in the south to the northern boundary of the Indo-Gangetic plain. Crustal shortening south of the normal faulting South Tibetan Detachment (STD; see Fig. 2.2B) resulted in major geological structures, namely, the Main Central Thrust (MCT) fault, the Main Boundary Thrust (MBT) fault, and the Main Frontal Thrust (MFT) fault. The MFT is the surface expression, at the front of the Siwalik hills, of the northward dipping thrust fault known as the MHT fault, where the Indian Plate subsides below the Eurasian Plate. The MFT lies between the sediments of the

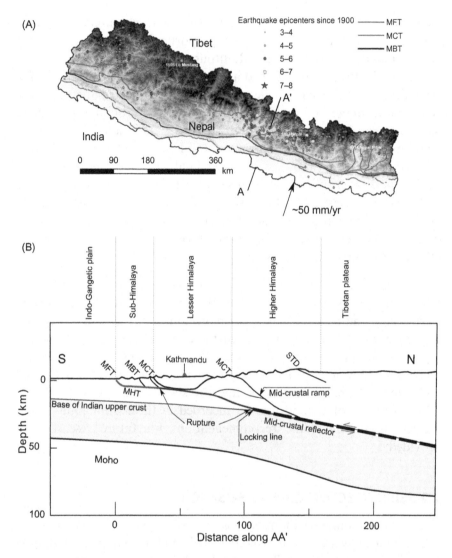

Figure 2.2 (A) Map of Nepal with the background representing digital elevation model (data provided by Jonathan de Ferranti, downloaded from http://www.viewfinderpanoramas.org/dem3.html). The circles indicate the epicenters of earthquakes larger than magnitude 3 since 1900, obtained from the NEIC catalog (https://earthquake.usgs.gov/earthquakes/search/). The epicenters of the April 24 main shock and May 12 aftershock are indicated by red stars (gray in print versions). (B) A schematic geological section across AA' (see (A)) based on Avouc (2003) and Bertinelli et al. (2006).

Indo-Gangetic plains and the molasses deposits of the sub-Himalaya (see Fig. 2.2B). The metasediments of the Lesser Himalaya are separated from the sub-Himalaya by the MBT. MCT separates Lesser Himalaya from the Higher Himalaya, which is mostly composed of gneisses and aplitic granite. A strong mid-crustal reflector exists at a depth of ~25 to ~40 km north of the Lesser Himalaya (see, for example, Zhao et al., 1993). This reflector has been interpreted as a ductile shear zone where the Indian Plate slides underneath the Eurasian Plate. South of this reflector, a shallow-dipping conductor coinciding with the surface between Indian and Eurasian plates has been identified (Lemonnier et al., 1999). This detachment surface, also known as the MHT, extends from the MFT to beneath southern Tibet, where it coincides with the mid-crustal reflector.

The present rate of 4−5 cm/year convergence of the two tectonic plates is accommodated partly by shear along the mid-crustal reflector and partly by crustal shortening (Larson et al., 1999). In central Nepal, the locked portion of the MHT (from the MFT to about 100 km north) absorbs about 2 cm of this convergence per year. Strain built up on the MHT is released during infrequent large earthquakes (see, for example, Bilham et al., 1995). The junction between the seismogenic and ductile zones defines a locking line. Intense microseismic activity occurs in a 50 km wide area around this locking line. Only about 0.2% of total accumulated seismic moment is released by the microseismic activity. The remaining energy is released during large earthquakes. According to Avouc (2003), only a part of the MHT, close to the down-dip end of the brittle zone, ruptures during earthquakes of magnitude around 7. During larger earthquakes, the MHT ruptures all the way to the MFT.

There has been only one instrumentally recorded earthquake with magnitude larger than 8 in Nepal: the 1943 magnitude 8.1 Nepal-Bihar Earthquake (Ambraseys and Douglas, 2004). The epicenter of this earthquake is shown in Fig. 2.2A. Historical records provide evidence of other large earthquakes in Nepal, such as the 1505 M 8.2 Lo Mustang Earthquake and the 1833 M 7.3 Nepal Earthquake. A massive earthquake in the Kathmandu Valley in 1255 is mentioned in Wright (1877) and is believed to have occurred in the same area as the 1934 Earthquake. Bollinger et al. (2016) provide some accounts of other historical earthquakes in Nepal. The earthquakes of 1934 and 1255 are believed to have ruptured all the way to the MFT. Other great earthquakes, in 1344 and 1408, are mentioned in historical chronicles (see Bollinger et al., 2016, for details) to have occurred in the same area as the 2015 Gorkha earthquake. The 1255 event is thought to have transferred stress to the west causing the 1344 event.

In a similar way, the 1934 event might have transferred stress to the west, causing the 2015 earthquakes. Unlike the 1934 event, the 2015 earthquake did not rupture to the MFT. Therefore, a significant part of strain accumulated on MHT south of Kathmandu has not yet been released. In addition to this, the area south and west of Kathmandu has not ruptured in a large earthquake since 1408. This unreleased strain might result in a large earthquake west and south of Kathmandu. Another possibility is that the unreleased strain is slowly being lost aseismically. Given the history of large earthquakes in the region rupturing all the way to the MFT, it would be wise to prepare for a larger earthquake, most likely to the south and west of Kathmandu in the near future. Although it is impossible to predict when the next large earthquake will occur, the sequence of earthquakes of 1255, 1344, and 1408 all happening within only a few decades calls for some caution. Regardless of when the next earthquake strikes, it is important to prepare for it by reducing social and infrastructure vulnerability to seismic risk. The seismic hazard in the region is difficult to quantify now, which necessitates more emphasis on reducing seismic vulnerability.

2.3 SOURCE MODELS AND THE RUPTURE PROCESS

The 2015 Gorkha Earthquake has been extensively studied using geodetic, teleseismic, and near-field seismic waveforms. No other large earthquake in the Himalayan arc has been studied in as great a detail as this earthquake. One of the earliest finite fault models of the main shock and the largest aftershock on May 12 was released by the United States Geological Survey (USGS). The slip distributions according to these models are shown in Fig. 2.3, along with the aftershocks reported by the NSC between April 24, 2015 and June 24, 2017. The USGS solution reports a strike of 295°, dip of 11°, and rake of 108°, and moment magnitude of 7.8 for the main shock. The USGS source model has a maximum slip of ~ 3 m, north of Kathmandu and about 70 km east of the hypocenter. The maximum slip estimated by USGS is significantly lower than that reported by Zhang et al. (2016), who report a maximum slip of 5.2 m based on inversion of teleseismic body waves. They report a moment magnitude of 7.9 and source duration of about 70 s. Wang and Fialko (2015) invert InSAR and GPS data to infer a moment magnitude of 7.8 and a maximum slip of 5.8 m at a depth of 8 km. Their source model has most of the slip in a

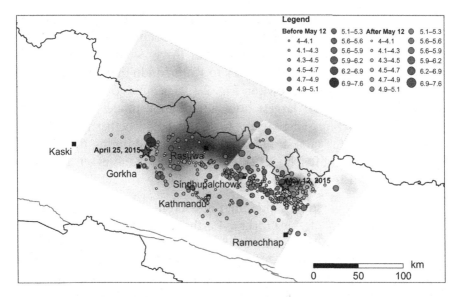

Figure 2.3 The source models of the 2015 Gorkha earthquake and its largest aftershock. The rectangles represent the fault plane solutions (according to the USGS) of the main shock and the May 12 aftershock. The colors represent slip ranging from 0 m (yellow, light gray in print versions) to 3 m (brown, black in print versions), and 0 to 4 m, for the main shock and aftershock, respectively. The circles indicate epicenters of earthquakes recorded by the NSC between April 24, 2015 and June 24, 2017, as indicated in the legend.

narrow zone between 50 and 100 km from the MFT at the deep end of the brittle zone of the MHT. Lindsey et al. (2015) use InSAR data to invert a finite fault model. The main shock slip was found to occur over an area ∼170 km long and between depths of 5 and 15 km. Maximum slip was estimated to be in the range of 5.5−6.5 m on a large asperity north of Kathmandu. Feng et al. (2015) use geodetic data to infer that the slip was confined at a depth of 7−15 km with largest slip occurring 15−20 km north of Kathmandu. Their results are in good agreement with the inversion of teleseismic waves reported by the USGS (shown in Fig. 2.3). Feng et al. (2015) estimate a maximum slip of 5.7 m and that both the main shock and the largest aftershock had a small right-lateral slip. They estimate that the maximum slip of the May 12, 2015 aftershock was 3.8 m. Similar to the USGS model, their model indicates an unruptured area between the main shock and the largest aftershock. The size of this area is 30−50 km, which could potentially generate an earthquake of moment magnitude close to 7. This is in contrary to the suggestion (Liu et al., 2016) that the risk of a large earthquake east of the epicenter of the main shock has decreased. Their (Liu et al., 2016) inversion of the main shock and the largest aftershock is based on seismic and geodetic data. Their model indicates a dominant thrust mechanism with a maximum slip of 5.8 m. The maximum slip of the aftershock was estimated to be ~ 5 m. The region of maximum slip in their model is closer to Kathmandu than in other models (for example, the one shown in Fig. 2.3). The areas of significant slip of the main shock and largest aftershock in the model of Liu et al. (2016) almost overlap. There seems to be a debate whether the ~ 20 km area just west of the largest aftershock has ruptured during the Gorkha Earthquake and its aftershocks (see, for example, Lindsey et al., 2015; Zhang et al., 2015).

Other numerous studies present finite fault slip model and high-frequency radiation of the Gorkha Earthquake (for example, Avouc et al., 2015; Fan and Shearer, 2015; Galetzka et al., 2015; Grandin et al., 2015; Meng et al., 2016; Wang and Mori, 2016; Yagi and Okuwaki, 2015). Most of the published results infer a large slip area east of the epicenter, with the high-frequency radiation occurring at the northern edge of the large slip area. The models, however, differ in terms of total slip, its distribution, and its maximum value. The models of Galetzka et al. (2015) and Lindsey et al. (2015) estimate significant slip near the surface, while most other models infer no significant slip near the surface. In lack of ground evidence of surface rupture, the consensus seems to be that the area south of Kathmandu has not ruptured and could potentially produce a large earthquake.

Qin and Yao (2017) present a very interesting analysis of the source process of the main shock using back projection of seismic array data. They divide the source radiation into six phases with time windows 0−8, 8−18, 18−28, 28−38, 38−47, and 47−56 s. The rupture propagated unilaterally to the east at an average speed of 2.8 km/s. The strongest radiation occured during the 28−38 s phase, by which time the rupture had traveled to a position about 30 km northeast of Kathmandu. They divide the rupture process into three stages (see Fig. 2.4). The first stage (0−18 s) is characterized by weak high-frequency radiation, few aftershocks, and a relatively small coseismic slip. The second stage (18−38 s) releases most of the energy. It is characterized by strong high-frequency radiation, relatively few aftershocks, and a large coseismic slip. The third stage (38−56 s) is the termination of rupture. It is characterized by weak high-frequency radiation, many aftershocks, and a relatively small coseismic slip. Qin and Yao (2017) also propose a multiscale asperity model, where the

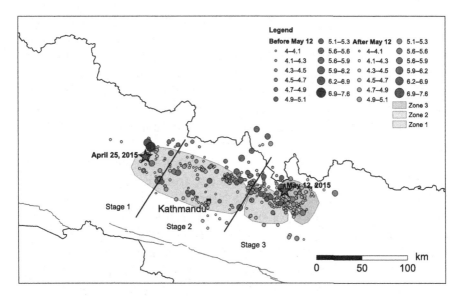

Figure 2.4 Multiple-scale asperity model (based on Qin and Yao, 2017), indicated by Zones 1, 2, and 3 of the main shock and the largest aftershock.

rupture area is divided into three zones. Zone 1, the eastern part, is characterized by a few large asperities fracturing during the main shock with a large coseismic slip. The middle part, Zone 2, contains a many small asperities. The eastern part, Zone 3, fractured during the May 12 aftershock. Their interpretation is that Zone 3 had accumulated relatively less stress than Zone 1, probably because some of the accumulated stress was released during the 1934 Earthquake. The stress heterogeneity caused the rupture to halt at the boundary of Zones 2 and 3. The stress change caused by rupture in Zones 1 and 2 ruptured Zone 3, 17 days later in an M$_W$ 7.3 event, which was followed by intense aftershock activity in this zone. Most of the areas with high-frequency radiations were located at the edges of large slip areas or at areas with small slip and dense aftershocks, indicating the role of small asperities in generating high-frequency radiation. Most of the high-frequency radiations were generated in the down dip area north of the high slip region.

2.4 SPATIOTEMPORAL DISTRIBUTION OF AFTERSHOCKS

The spatial and temporal distribution of aftershocks of the Gorkha Earthquake have been described in many studies, such as Baillard et al. (2017), Ichiyanagi et al. (2016), Adhikari et al. (2015), and McNamara et al. (2016). In terms of magnitude-frequency distribution, the smaller aftershocks (less than moment magnitude ~ 6) are found to follow a linear Gutenberg-Richter model (see, for example, Rupakhety et al., 2017). Using aftershock data from April 25, 2015 to June 21, 2016, Rupakhety et al. (2017) report a least-squares estimate of b-value equal to 0.85. Adhikari et al. (2015) use data (completeness at a local magnitude of 4) recorded at stations in Nepal within 45 days of the main shock and report a b-value of 0.8 with a standard error of 0.05. They report that the b-value estimated from earthquakes recorded between 1995 and 2015 with events larger than a local magnitude of 2 is 0.83 with a standard error of 0.05: an estimate close to the b-value of the Gorkha Earthquake aftershocks. Baillard et al. (2017) use automatically identified events from the local catalog and report a b-value of 0.92 for a 3-month period after the main shock. Frequency distribution of earthquakes between the main shock and February 27, 2017, reported in the NEIC catalog is shown in Fig. 2.5. The Gutenberg-Richter model fitted to the data using the maximum likelihood method is shown in the figure by the dashed black line. The estimated model is almost the same as the one reported in Baillard et al. (2017), who used

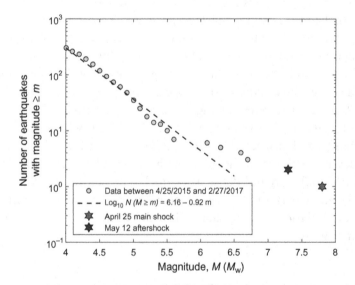

Figure 2.5 Magnitude-frequency distribution of earthquakes larger than magnitude 4 in the region of the 2015 Gorkha Earthquake sequence. The earthquakes are taken from the NEIC catalog between the main shock and February 27, 2017. The dashed line represents the Gutenberg-Richter model obtained from a maximum likelihood method. The standard error of the b-value is 0.05.

data recorded by stations located in Nepal. The results indicate that, while a linear relationship is valid for earthquakes up to magnitude ~6, larger earthquakes seem to be associated with smaller b-values. The slope of the line following earthquakes between magnitude 6 and 7.8 in Fig. 2.5 is almost half the slope of the line following the smaller earthquakes.

The spatial distribution of aftershocks of the 2015 Gorkha Earthquake shows some interesting features. As can be inferred from Fig. 2.3, most of the aftershocks occurred next to areas of largest coseismic slip. Within 12 h after the main shock, more than 100 after-shocks with local magnitude larger than 4 occurred in a wide area extending as far as the location of the largest aftershock, which occurred on May 12, 2015. Rate of aftershocks decayed in the following days, until the M$_w$ 7.3 aftershock on May 12, which was followed by many other aftershocks. The events after the M$_w$ 7.3 aftershock were clustered around its epicenter. Significant aftershocks continued in the following days in the western part of the fault (near the main shock) and in the area between the main shock and the largest after-shock. McNamara et al. (2016) find that the depths of the aftershocks are consistent with the MHT. Ichiyanagi et al. (2016), however, report that most aftershocks near Kathmandu occurred in the upper crust, at depths shallower than 10 km. Bai et al. (2017) also argue that most of the aftershocks occurred above the MHT on the hanging wall. They infer that the aftershocks are distributed above the anticlinorium system of the MCT, which are steeper than the MHT.

The temporal distribution of the aftershocks also shows some interesting features. To study the temporal distribution, the same NEIC catalog as that used in Fig. 2.5 is analyzed. Fig. 2.6 shows the cumulative number of earthquakes (larger than or equal to M$_w$ 4.5) as a function of number of days after the main shock. In Fig. 2.6A, the cumulative number of earthquakes within 3 days of the main shock is shown by the red curve. The blue curve represents the Omori-Utsu model fitted to the registered number of earthquakes in this time span. The corresponding model parameters are shown in the figure legend; the P-value, which represents the rate of decay of earthquakes, is equal to 1.11. The model seems to fit the observations well in the early hours after the earthquake, but a relatively quiet period is observed after about 1.5 days following the main shock. To illustrate this quiet period better, in Fig. 2.6B, the model fitted to the observations in the first

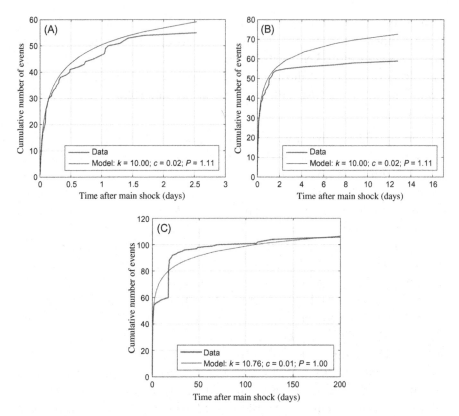

Figure 2.6 Comparison of the cumulative number of aftershocks as a function of days passed after the main shock with the Omori-Utsu model using NEIC catalog between April 25, 2015 and February 27, 2017, and earthquakes of size larger than magnitude 4.5. In (A) the activity in the first 3 days after the main shock is compared with the corresponding Omori-Utsu model. In (B) the activity before the largest aftershock, on May 12, is compared with the model fitted to the data in the first 3 days of the main shock. In (C) activity in the 200 days following the earthquake is compared to the corresponding model.

3 days is compared to the observed number of earthquakes before the largest aftershock of May 12, 2017. It can be observed that the rate of earthquakes decreased dramatically after about 1.5 days following the main shock, an observation also made by Rupakhety et al. (2017) and Ogata and Tsuruoka (2016). The quite period lasted for almost 2 weeks, when a large aftershock of M_w 7.3 occurred. The occurrence of this aftershock triggered other secondary aftershocks, and the cumulative number of earthquakes increased drastically. This increase is evident in Fig. 2.6C, which shows the observed number earthquakes 200 days after the main shock and the Omori-Utsu model fitted to it. Comparison of Fig. 2.6A and C shows that the Omori-Utsu model fitted to observations in the first 3 days of the main shock is quite similar

to the one fitted to observations 200 days later. However, the observed number of earthquakes between days 2 and 17 of the main shock was much lower than what would be predicted by Omori-Utsu law. This phenomenon has been interpreted by Ogata and Tsuruoka (2016) as seismic quiescence. They suggest that lower aftershock activity than expected can be an indicator of increased probability of a large aftershock, a feature that is also described in Ogata (2001). In hindsight, their argument is valid for the Gorkha Earthquake and its largest aftershock.

2.5 GROUND MOTION CHARACTERISTICS

The 2015 Gorkha Earthquake is, unfortunately, a missed opportunity to record strong ground motion due to large earthquakes in the region. Only a few strong-motion stations were operational at the time of the main shock. The station KATNP (see Fig. 2.7) operated by the USGS in Kathmandu recorded the main shock and many aftershocks. Ground motion records from this station were released soon after the earthquake and played an important role in understanding the salient features of ground shaking in Kathmandu. In addition, the station

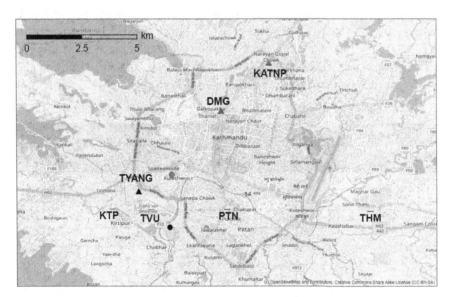

Figure 2.7 Map of the Kathmandu valley and surrounding area showing the locations of the strong-motion stations that recorded the Gorkha earthquake and its aftershocks. The dots *indicate the locations where deep borehole logs are available. The map data is obtained from OpenStreetMap (openstreetmap.org) released under Open Data Commons Open Database License (ODbL).*

operated by the DMG (see Fig. 2.7) recorded the main shock and many aftershocks. Although some figures showing ground motion traces recorded at the station were released, ground motion data was withheld for a long time before being released as an electronic supplement to Bhattarai et al. (2015). Ground motion during the main shock at four other stations (KTM, TVU, PTN, and THM in Fig. 2.7) in Kathmandu Valley was released by Takai et al. (2016). Ground motion of some aftershocks recorded at station TYANG (see Fig. 2.7), operated in Kathmandu by the Earthquake Engineering Research Center of the University of Iceland, has also been released as an electronic supplement to Rupakhety et al. (2017). Of these six stations, KTP lies on a local outcrop of the bedrock, while the others lie on sediments of the valley.

An overview of the ground motion recorded at the DMG station is provided in Bhattarai et al. (2015). They report amplification of waves in horizontal direction at a frequency of 0.25 and 0.3 Hz for the main shock and aftershock, respectively. Rupakhety et al. (2017) present a detailed analysis of strong ground motion in the Kathmandu Valley. Some of the important results of the analysis are discussed here. The peak ground acceleration (PGA) and root-mean-squared (RMS) acceleration recorded at the six strong-motion stations in the Kathmandu Valley are presented in Table 2.1. The horizontal PGA varies from 0.13 to 0.26 g (acceleration due to gravity), which is considered low for an earthquake of this size (see also Dhakal et al., 2016). Although the PGA in the vertical direction is larger than that in the horizontal

Table 2.1 Peak ground acceleration, RMS ground acceleration, and Peak Factors of the ground motion recorded at six stations in the Kathmandu Valley during the main shock

Station	PGA (g)			RMS (g)			PF		
	EW	NS	V	EW	NS	V	EW	NS	V
KATNP	0.16	0.16	0.19	0.04	0.04	0.03	3.74	3.71	5.72
THM	0.14	0.15	0.19	0.05	0.05	0.03	2.95	3.00	6.20
PTN	0.13	0.15	0.14	0.04	0.04	0.03	2.55	3.77	4.55
TVU	0.23	0.21	0.14	0.06	0.05	0.03	4.05	4.01	5.56
KTP	0.26	0.16	0.13	0.04	0.03	0.02	7.12	5.21	5.31
DMG	0.13	0.18	0.21	0.04	0.04	0.04	3.31	4.86	4.93

The RMS acceleration is computed over a time window covering 5%–95% of the rotation-invariant Arias Intensity (Rupakhety and Sigbjörnsson, 2014).

direction at many of the stations located in the valley (KATNP and DMG, for example), the vertical motion is consistently less energetic, as is indicated by lower RMS acceleration values, except at the DMG station. Despite higher RMS acceleration, peak acceleration in the horizontal direction is in general lower than that in the vertical direction, which indicates large peak factors, indicative of high-frequency content in the vertical motion, except at the rock site KTP, where the horizontal motion contains stronger high-frequency waves. High-frequency motion was likely filtered out by the soft sediments of the valley, resulting in lower peak ground acceleration than those at stiffer sites.

In terms of total energy content quantified by Arias Intensity (AI), Rupakhety et al. (2017) found that horizontal motion at KATNP had much larger energy content than vertical motion. In contrast, the vertical motion at DMG has larger AI than horizontal motion. Rupakhety et al. (2017) also compared the AI at the first five stations listed in Table 2.1. It was observed that the motion at the rock site KTP, despite its high PGA, had the least AI. The motion at TVU was peculiar with significantly larger AI than the other stations (about three times larger than that at the nearby station KTP). The motion at TVU is characterized by several large amplitude pulses, which would be more severe to structures deformed beyond their elastic limit than the motion at KTP, which despite the larger PGA, is characterized by strong motion in relatively few cycles. It was found that the contribution of a pulse with period close to 5 s to the AI is relatively large at softer sites but low at the rock site. The total AI of the motion filtered to remove the frequencies close to 5 s were similar at the stations except at TVU. These results indicate that the difference in the AI at the stations (except TVU) is mainly in the frequency band near 0.2 Hz. In terms of dominant frequencies, Rupakhety et al. (2017) report that the rock site KTP has a relatively high dominant frequency at 2.5 Hz, while KATNP, THM, and PTN have much lower dominant frequency of 0.2 Hz.

Elastic response spectra of the main shock motion recorded at KATNP is presented in Fig. 2.8. The spectra are computed for uncorrelated horizontal motion, also known as principal components. The rotation-invariant response spectra (Rupakhety and Sigbjörnsson, 2016) are also shown in the figure. A very conspicuous peak at a period close to 5 s is observed in the response spectra. This peak is

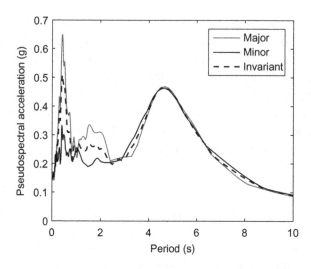

Figure 2.8 Elastic response spectra (5% damped) of the main shock motion recorded at KATNP.

similar for motion in both the major and minor principal directions, although quite different for correlated motion in the east-west and north-south directions. This peak has been interpreted as the contribution of a ~5 s pulse that was dominant in the ground motion. Galetzka et al. (2015) interpret the 5 s pulse as a resonance of the valley. Dhakal et al. (2016), however, conclude that the 5 s pulse can be well explained by a one-dimensional site response. It has been argued that the concentration of energy in the long-period range spared regular dwellings (which have relatively short vibration periods) in the Kathmandu Valley but caused significant damage to tall structures. While there is some merit to this argument, a closer look at the response spectra at KATNP (see Fig. 2.8) raises other questions. While much attention has been given to the peak of the spectra between 4 and 6 s, the larger peak around 0.3 s seems to be largely ignored by many analysts. This peak has an amplitude of about 0.65 g. As a rough rule of thumb, a fundamental vibration period of 0.3 s corresponds to a three-story reinforced-concrete frame building with masonry infill walls, which are very common in Kathmandu. These buildings experienced significant demand during the earthquake, with base shear coefficient as high as 0.6 g. At the same time, it is debatable whether the long-period pulse of 4–5 s period caused collapse of tall structures. A 4 s period would correspond to roughly 40-story building, which did not exist in Kathmandu. A closer look at the response spectra shown in Fig. 2.8 shows that, after the high peak around 0.3 s, the plateau of the spectra is very wide, extending almost up to 2 s. At such periods, the Eurocode 8 (EC8) spectra, for example, predicts spectral acceleration much lower than the PGA. Whereas in this case, the spectral acceleration up to 2 s is almost twice the peak ground acceleration. It is therefore clear that despite the low PGA, the spectral acceleration was rather high for periods up to 2 s, which is the spectral range of most structures in Kathmandu. It is possible, as argued by some (Dhakal et al., 2016, for example), that nonlinear deformation of the soil in the valley might have reduced the PGA and amplified long-period waves.

Rupakhety et al. (2017) compare the average normalized response spectra of the main shock and large aftershocks recorded at KATNP with the EC8 site Type 1 spectrum for site class D. The comparison shows that the EC8 spectral shape captures the overall feature of the recorded spectral shape up to a period of about 1 s. It, however, fails to model the peak of the recorded spectral shape, which is around 3.5 as opposed to 2.5 in the EC8 model. Chaulagain et al. (2015) present 475-year uniform hazard spectra (UHS) for the study area based on probabilistic seismic hazard assessment (PSHA). A direct comparison of the UHS with observed motion is not fair. However, when normalized by the corresponding PGA, the spectral shape should be capable of modeling the observed spectral shape. The spectral shape inferred from Chaulagain et al. (2015) is very different from the average spectral shape of the main shock and aftershocks recorded at KATNP. This underlines a fundamental problem in PSHA: use of ground motion prediction models that are not suitable for the region.

2.5.1 Other Source and Local Effects

Source effects played an important role in the spatial distribution of ground shaking and consequently damage caused by the Gorkha Earthquake. Eastward propagation of rupture caused extensive damage to areas east of the hypocenter. The asymmetric distribution of damage about the hypocenter is clearly visible in Fig. 2.1. It seems that the rupture, which started in Gorkha and propagated eastward, slowed down at a transition zone between high- and low-velocity areas and eventually halted in Zone 2 (see Fig. 2.4). This resulted in stress build-up in Zone 3, causing it to rupture with an M_w 7.3 event on May 12 and intense aftershock activity afterward. This aftershock caused extensive damage in the districts north and east of Kathmandu and collapsed many buildings that were partially damaged by the main shock. The onset of rupture and its early stage occurred on a large-scale asperity (Qin and Yao, 2017) near the hypocenter, with smooth slip onset (Galetzka et al., 2015) and relatively stable rupture propagation and resulted in moderate high-frequency radiation, which were mostly concentrated on the down-dip part of the MHT at many small-scale asperities. This source effect was responsible for relatively low high-frequency motion and partly responsible for strong low-frequency shaking in Kathmandu.

It has been argued that forward directivity effect generated long-period motion east of the hypocenter. Koketsu et al. (2016) argue that the strong pulses observed in the initial phases of ground motion recorded in Kathmandu are a combination of forward directivity and permanent displacement (fling) pulse. Forward directivity pulses are aligned on the fault-normal component and occur at sites toward which the rupture propagates. The Gorkha earthquake had rupture propagation in the strike direction. Koketsu et al. (2016) argue that, for low-dip events such as the Gorkha Earthquake, forward directivity pulses can be generated at the ground surface above the slipping fault even when the rupture propagation is along the strike. They also argue that the reason for more damage to the east is due to the forward directivity pulses. There is no doubt about the directionality of ground shaking: It was stronger in the east. The reason for this is not necessarily forward directivity pulses. Most of the rupture on the fault occurred east of the hypocenter and areas above the largest slips experienced most damage. In addition, the May 12 aftershock caused further extension of damage area to the east. Koketsu et al. (2016) perform numerical simulations to demonstrate generation of pulselike ground motion, which they classify as fling pulse and directivity pulse. What they refer to as *simulated directivity pulse* are mostly one-sided velocity pulses, which indicate permanent displacement. Directivity pulses are expected to be two sided. The period of the simulated pulse is quite long, about 5 s. Such long-period pulse should not be critical for most of the structures in the damaged area, which have much higher vibration frequencies. The argument that the damage distribution implies a forward directivity pulse, therefore, does not seem too convincing. The damage distribution is explained, as mentioned earlier, by the areas of largest slip on the fault plane. Forward directivity pulses are thought to be aligned on the fault-normal component, which in this case would be the vertical motion. Kokeksu et al. (2016) infer that a forward directivity pulse was present also in the north-south motion in the Kathmandu valley.

Zare et al. (2017) indicate that forward directivity pulses were present in the vertical component of motion at KATNP and attribute a frequency of 1.5−3 Hz (period of 0.33−0.67 s), which they claim are also visible on the velocity and displacement time histories of the KATNP motion. The velocity and displacement waveforms of the KATNP motion obtained by Rupakhety et al. (2017) are shown in Fig. 2.9. The traces do not contain any pulses with periods inferred by Zare et al. (2017). The pulses are of much longer period, around 5 s. The initial pulse in the horizontal direction is one sided causing a permanent displacement to the west and south: These can be interpreted as fling pulses. In the vertical direction, the velocity pulse is very much one sided and lack the oscillatory nature present in the horizontal motion. This one-sided pulse is a fling effect rather than a directivity effect. Takai et al. (2016), using data at both the rock site and KATNP station, conclude that the velocity pulses were due to permanent displacement (fling effect). For example, the north-south velocity at the rock site KTP, as reported in Takai et al. (2016), shows a very clear one-sided pulse and lacks the oscillatory pulses, which were

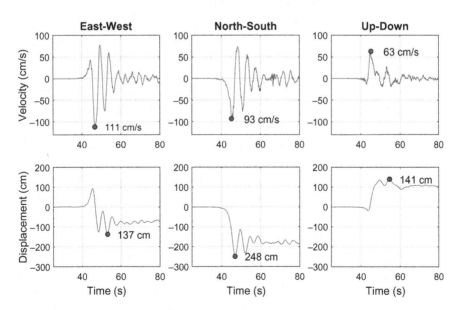

Figure 2.9 Ground velocity (top panel) and displacement (bottom panel) obtained from ground acceleration recorded at KATNP during the April 24 main shock.

observed at the sedimentary sites such as KATNP. This is a clear evidence that the interpretation of Koketsu et al. (2016) that the first pulse observed at DMG is due to fling effect and the second pulse is due to a directivity effect is not convincing. The oscillatory pulses observed in the valley were most likely due to site effects, an inference that is further supported by the lack of oscillatory pulses in the vertical motion. It is also interesting to note that the first pulse in the north-south component at DMG, as reported in Koketsu et al. (2016), is of very small amplitude, not capable of producing the permanent displacement confirmed in other studies (Rupakhety et al., 2017; Takai et al., 2016; Galetzka et al., 2015). This is probably due to the signal processing algorithm used to derive the velocity time history from the recorded acceleration time history. In summary, it can be said that the pulses observed in ground motion recorded in Kathmandu were mainly due to permanent displacement effect rather than forward directivity effect. The frequency of the pulse was such that it caused resonance in the sediments of the valley, which resulted in oscillatory pulses in the horizontal motion at soft sites, a feature lacking at rock sites, as well as vertical motion at soft sites.

2.5.2 Topographic and Basin Effects

Some studies indicate that a basin-edge effect and hilltop-edge effect dominated the distribution of ground failure and structural damage distribution caused by the Gorkha Earthquake sequence. Hashash et al. (2015) and Wang et al. (2016) mention some examples of potentially stronger ground shaking and greater damages at basin edges and hilltops. Some examples of such locations are Bhaktapur, Sankhu, Siddhitol, Ramkot, and Swayambhunath Temple. In Kirtipur, which lies on the edge of the basin, even stone masonry buildings with mud mortar survived the earthquake without significant damage. It has been suggested that lower damage observed in Kirtipur is due to the shallow depth of bedrock. While it is tempting to make conclusions about different site effects based on observed damage, in lack of recorded data, it is important to consider factors such as building vulnerability and other structural deficiencies. The author observed many examples of heavily damaged buildings built on sloping ground. While these examples could be interpreted as topographic effect, structural deficiencies caused by columns supported at different levels makes them vulnerable to ground shaking.

Hashash et al. (2015) interpret the long-period ground motion waves observed in the Kathmandu valley as a two-dimensional basin effect, in particular, a creation of Rayleigh waves similar to the one described in Bard and Bouchon (1985). Taking valley depth as 250 m and half-width of 5 km, the shape factor of the Kathmandu valley is 0.05, which according to Bard and Bouchon (1985) is a shallow valley where the fundamental frequency of valley resonance is equal to the resonance frequency of flat layers. For these types of valleys, one-dimensional resonance with lateral propagation of surface waves is expected to dominate the valley response. The polarization ellipsoid of the long-period pulse at KATNP is nearly horizontal (see Rupakhety et al., 2017), which does not correspond to the polarization of Rayleigh waves. The polarization pattern is similar to those of Love waves. As reported in Rupakhety et al. (2017), the Fourier amplitude of vertical ground motion at KATNP is much lower than that at KTP, at a frequency of around 0.2 Hz. The polarization of the ~ 0.2 Hz pulse on the horizontal plane is potentially responsible for this. The spectral content at 0.2 Hz is significant at the rock site KTP. This indicates that the source radiated low-frequency motion, which was further amplified by the sediments of the valley. Nonlinear deformation of the sediments in the basin might have caused lower high-frequency motion in the valley compared to that at KTP.

2.6 CONCLUSIONS

The 2015 Gorkha Earthquake sequence was the worst disaster in Nepal since the 1934 Earthquake. The sequence started with an M_w 7.8 earthquake that ruptured the locked part of the MHT below central Nepal. The fault is of thrust type, dipping slightly ($7° - 10°$) to the north. The source models inferred from teleseismic and geodetic data reveal a very complex process. The rupture started west of Kathmandu and radiated relatively low-frequency waves. The rupture then propagated eastward along its strike at a steady velocity. The initial rupture area contained a few large asperities that radiated low-frequency motion. The rupture slowed down and eventually terminated east of Kathmandu. The main shock was followed by several large aftershocks, the largest one with M_w 7.3 occurring 17 days later in an area to the east of the main shock rupture area. The strongest radiation and those with high frequency occurred near the down-dip edge of the fault. Aftershocks occurred mainly at the edges of areas with the largest slips. After 2 days of intense aftershock activity, the number of aftershocks decreased significantly for 15 days, then rapidly increased again after the May 12 M_w 7.3 event. Strong ground motion caused by the main shock and aftershock were recorded by only a few accelerometers located in Kathmandu. Ground motion in the valley, which is filled with lacustrine deposits, was dominated by a strong pulse of frequency ~ 0.2 Hz. This long-period pulse was most likely caused by the source effect (smooth onset of rupture and relatively long rise time) and amplified by the soft sediment of the valley. The presence of a relatively large amplitude of motion at 0.2 Hz frequency at the rock site KTP indicates that the long-period pulse was not only due to a site effect. The motion recorded at soft sites had lower high-frequency content than the motion recorded at the rock site. Nonlinear deformation of the sediment might resulted in a filtering of high-frequency waves.

The nature of ground motion, and especially the shape of their response spectrum, is very poorly represented by empirical ground motion prediction equations (GMPEs) and spectral shapes specified in design codes. A direct implication of this is that UHS obtained from PSHA using ground motion models calibrated with data recorded in other tectonic environments are not reliable. Ground motion models for subduction zone earthquakes are mostly based on data recorded from subduction events at oceanic ridges. The MHT is a continental subduction fault, and the rupture area is much shallower than most of the oceanic subduction events. The source effects, as discussed, were also peculiar and resulted in long-period ground motion. Such effects cannot be expected to be modeled by empirical GMPEs. In this context, PSHA for seismic design in Kathmandu and the surrounding area is very challenging.

The Gorkha Earthquake and its aftershocks were very tragic but also a lost opportunity. While the geodetic measurements in this region have been exceptionally good, strong-motion measurement, which provides data relevant for engineering design, has been extremely poor. The DMG of the government of Nepal has been operating a network of seismometers for many years but seems to have largely overlooked the need for a strong-motion measurement system. Had a network of accelerometers been operational, the Gorkha Earthquake sequence would have provided valuable data that could have been used to calibrate local GMPEs or at least constrain GMPEs that have been calibrated elsewhere. This missed opportunity needs to be taken as a lesson. Since the earthquake did not rupture the part of the locked fault south of Kathmandu and, according to some studies, a wide area between Kathmandu and the largest aftershock, strong earthquakes can be expected in the area in the near future. A strong earthquake to the south or west of Kathmandu cannot be ruled out, although it is impossible to predict where and when such an earthquake will strike. Despite of their unpredictability, earthquakes are a continuing threat in Nepal.

The Gorkha earthquake could have been a lot more devastating. First, it occurred on a Saturday noon, when children were not in school. The number of completely damaged classrooms and collapsed buildings indicates that number of casualties could have been much higher if the earthquake had struck at night or on a school day. In addition, the earthquake radiated more low-frequency waves than high-frequency ones. The filtering of high-frequency waves by the soft sediment of the Kathmandu Valley likely reduced demand on most structures in the valley. A different size earthquake, even a smaller one radiating more low-frequency waves, could have been more damaging to buildings in the valley. The complex nature of source and site effects is still poorly understood due to a lack of reliable geotechnical information. The lack of recorded ground motion data over a large spatial area has further hindered research and understanding of the ground shaking characteristics.

It is frustrating that the government and its responsible body, DMG, 2 years after the earthquakes, has not taken any concrete steps to improve strong-motion monitoring system in Nepal. There have been some efforts from the nongovernment sector to establish a network of low-cost microelectromechanical system sensors. Although this effort is a positive contribution, engineering seismologists need to focus on establishing a dense network of high-quality accelerometers all over the country. There seems to be more interest and expertise on research related to geology and geoscience than engineering seismology in Nepal. The DMG and its National Seismological Centre seem to lack resources and support from the government to make a big leap in engineering seismological research in Nepal. A proper strong-motion network funded by the government is urgently needed. In addition, dense small-aperture arrays should be installed at some locations in Kathmandu Valley. An infrastructure as important as this should not be delegated to nongovernment organizations or volunteers. One potential way forward is for the DMG to continue its work on earthquake location and monitoring and for the government to establish a separate department, either within its ministries or preferably as an independent entity of engineers and researchers, to pursue strong-motion monitoring and engineering seismological research. The institute of engineering of Tribhuwan University should take a leading role in such research and developments.

Another pressing issue that needs to be addressed urgently is seismic microzonation of Kathmandu Valley. Some attempts have been made in the past by using microtremor measurements. Such quick methods can be useful as a preliminary study, but ground motion records from Kathmandu Valley show that predominant frequencies inferred from microtremor measurements are not very reliable. In this regard, geotechnical investigation with boreholes and experimental validation of soil properties needs to be done. In the lack of a proper monitoring system, local data, and reliable geotechnical information, seismic hazard estimates in Kathmandu will remain uncertain. Irrespective of the sophistication of computational models and tools used in PSHA, lack of knowledge of local source, path, and site parameters makes the results of such analysis unreliable. In lack of recorded ground motion data, physics-based ground motion simulation may be used to estimate some bounds on expected hazards. Even for such applications, path and site parameters need to be estimated accurately. In addition to advancement of research in engineering seismology, equal emphasis should be placed on reducing seismic vulnerability of existing and future constructions.

ACKNOWLEDGMENT

This work was partially funded by a grant from the University of Iceland Research Fund.

REFERENCES

Adhikari, L.B., Gautam, U.P., Koirala, B.P., Bhattarai, M., Kandel, T., Gupta, R.M., et al., 2015. The aftershock sequence of the 2015 April 25 Gorkha–Nepal earthquake. Geophys. J. Int. 203, 2119–2124.

Ambraseys, N.N., Douglas, J., 2004. Magnitude calibration of North Indian earthquakes. Geophys. J. Int. 159, 165–206.

Avouc, J.P., 2003. Mountain building, erosion, and the seismic cycle in the Nepal Himalaya. Adv. Geophys. 46, 1–80.

Avouc, J.P., Meng, L., Wei, S., Want, T., Ampuero, J.P., 2015. Lower edge of locked main Himalayan Thrust unzipped by the 2015 Gorkha earthquake. Nat. Geosci. 8, 708–711.

Bai, L., Liu, H., Ritsema, J., Mori, J., Zhang, T., Ishikawa, Y., et al., 2017. Faulting structure above the Main Himalayan Thrust as shown by relocated aftershocks of the 2015 Mw 7. 8 Gorkha, Nepal, Earthquake. Geophys. Res. Lett. 43 (2), 637–642.

Baillard, C., Lyon-Caen, H., Bollinger, L., Rietbrock, A., Letort, J., Adhikari, L.B., 2017. Automatic analysis of the Gorkha earthquake aftershock sequence: evidences of structurally segmented seismicity. Geophys. J. Int. 209 (2), 1111–1125.

Bard, P.Y., Bouchon, M., 1985. The two-dimensional resonance of sediment-filled valleys. Bull. Seismol. Soc. Am. 75 (2), 519–541.

Bertinelli, P., Avouc, J.-P., Flouzat, M., Jouanne, F., Bollinger, L., Willis, P., et al., 2006. Plate motion of India and interseismic strain in the Nepal Himalaya from GPS and DORIS measurements. J. Geodyn. 80, 567–589.

Bhattarai, M., Adhikari, L.B., Gautam, U.P., Laurendeau, A., Labonne, C., Hoste-Colomer, R., et al., 2015. Overview of the large 25 April 2015 Gorkha Nepal, earthquake from accelerometric perspectives. Seismol. Res. Lett. 86 (6), 1540–1548.

Bilham, R., Bodin, P., Jackson, M., 1995. Entertaining a great earthquake in Western Nepal: historic activity and geodetic test for the development of strain. J. Nepal Geol. Soc. 11, 73–88.

Bollinger, L., Tapponnier, P., Sapkota, S.N., Klinger, Y., 2016. Slip deficit in central Nepal: Omen for a repeat of the 1344 AD earthquake? Earth Planets Space 68, 12.

Chaulagain, H., Rodrigues, H., Silva, V., Spacone, E., Varum, H., 2015. Seismic risk assessment and hazard mapping in Nepal. Nat. Hazards 78 (1), 583–602.

Dhakal, Y.P., Kubo, H., Suzuki, W., Kunugi, T., Aoi, S., Fujiwara, H., 2016. Analysis of strong ground motions and site effects at Kantipath, Kathmandu, from 2015 Mw 7.8 Gorkha, Nepal, earthquake and its aftershocks. Earth Planets Space 68 (1), 58.

Ekström, G., Nettle, M., Cziewonski, A.M., 2012. The global CMT Project 2004–2010: centroid-moment tensors for 13,107 earthquakes. Phys. Earth Planet Inter. 200–201, 1–9.

Fan, W.Y., Shearer, P.M., 2015. Detailed rupture imaging of the 25 April 2015 Nepal earthquake using teleseismic P waves. Geophys. Res. Lett. 42, 5744–5752.

Feng, G., Li, Z., Shan, X., Zhang, L., Zhang, G., Zhu, J., 2015. Geodetic model of the 2015 April 25 Mw 7.8 Gorkha Nepal earthquake and Mw 7.3 aftershock estimated from InSAR and GPS data. Geophys. J. Int. 203, 896–900.

Galetzka, J., Melgar, D., Genrich, J.F., Owen, J., Lindsey, S., Xu, X., et al., 2015. Slip pulse and resonance of the Kathmandu Basin during the 2015 Gorkha Earthquake, Nepal. Science 349, 1091–1095.

Grandin, R., Vallee, M., Satriano, C., Lacassin, R., Klinger, Y., Simoes, B., et al., 2015. Rupture process of the Mw = 7.9 2015 Gorkha Earthquake (Nepal): insights into Himalayan Megathrust imaged by the Hi-net array. Nature 435, 933–936.

Hashash YM, Tiwari B, Moss RE, Asimaki D, Clahan KB, Kieffer DS, et al., 2015. Geotechnical field reconnaissance: Gorkha (Nepal) earthquake of April 25, 2015 and related shaking sequence. Geotechnical Extreme Event Reconnaisance GEER Association Report No. GEER-040, 2015:1–250.

Ichiyanagi, M., Takai, N., Shigefuji, M., Bijukchhen, S., Sasatani, T., Rajaure, S., et al., 2016. Aftershock activity of the 2015 Gorkha, Nepal, earthquake determined using the Kathmandu strong motion seismographic array. Earth Planets Space, 68 (1), 25.

Koketsu, K., Miyake, H., Guo, Y., Kobayashi, H., Masuda, T., Davuluri, S., et al., 2016. Widespread ground motion distribution caused by rupture directivity during the 2015 Gorkha, Nepal earthquake. Sci. Reports 6, 28536.

Larson, K.M., Burgman, R., Bilham, R., Freymueller, J.T., 1999. Kinematics of the India–Eurasia collision zone from GPS measurements. J. Geophys. Res. 104, 1077–1093.

Lemonnier, C., Marquis, G., Perrier, F., Acouc, J.-P., Chitrakar, G., Kafle, B., et al., 1999. Electrical structure of the Himalaya of central Nepal: high conductivity around the midcrustal ramp along the MHT. Geophys. Res. Lett. 26 (21), 3261–3264.

Lindsey, E.O., Natsuaki, R., Xiaohua, Xu, Shimada, M., Hashimoto, M., Melgar, D., et al., 2015. Line-of-Sight Displacement from ALOS-2 Interferometry: Mw 7.8 Gorkha Earthquake and Mw 7.3 Aftershock. Geophys. Res. Lett. 42 (16), 1655.

Liu, C., Zheng, Y., Wang, R., Shan, B., Xie, Z., Xiong, X., et al., 2016. Rupture process of the 2015 Mw 7.9 Gorkha earthquake and its Mw aftershock and their implications on the seismic risk. Tectonophysics 682, 264–277.

McNamara, D.E., Yeck, W.L., Barnhart, W.D., Schulte-Pelkum, V., Bergman, E., Adhikari, L. B., et al., 2016. Source modeling of the 2015 Mw 7.8 Nepal (Gorkha) earthquake sequence: implications for geodynamics and earthquake hazards. Tectonophysics. Available from: http://dx.doi. org/10.2016/j.tecto.2016.08.004.

Meng, L., Zhang, A., Yagi, Y., 2016. Improving back projection imaging with a novel physics-based aftershock calibration approach: a case study of the 2015 Gorkha earthquake. Geophys. Res. Lett. 43, 628–636.

National Seismological Centre (NSC), 2017. <http://www.seismonepal.gov.np/> (accessed 20.06.17.).

Nepal Disaster Risk Reduction Portal (NDRRIP), 2015. Nepal earthquake: disaster relief and recovery information platform. <http://drrportal.gov.np/> (accessed 06.06.17.).

Ogata, Y., 2001. Increased probability of large earthquakes near aftershock regions with relative quiescene. J. Geophys. Res. 106, 8729–8744.

Ogata, Y., Tsuruoka, H., 2016. Statistical monitoring of aftershock sequences: a case study of the 2015 Mw7.8 Gorkha, Nepal, Earthquake. Earth Planets Space. Available from: http://dx.doi.org/ 10.1186/s40623-0.16-0410-8.

Qin, W., Yao, H., 2017. Characteristics of subevents and three-stage rupture process of the 2015 Mw 7.8 Gorkha Nepal earthquake from multiple-array back projection. J. Asian Earth Sci. 133, 72–79.

Rupakhety, R., Sigbjörnsson, R., 2014. Rotation-invariant mean duration of strong ground motion. Bull. Earthquake Eng. 12 (2), 573–584.

Rupakhety, R., Sigbjörnsson, R., 2016. Rotation-invariant measures of earthquake response spectra. Bull. Earthquake Eng. 11 (6), 1885–1893.

Rupakhety, R., Olafsson, S., Halldorsson, B., 2017. The 2015 Mw 7.8 Gorkha earthquake in Nepal and its aftershocks: analysis of strong ground motion. Bull. Earthquake Eng. 15 (7), 2587–2616.

Takai, N., Shigefuji, M., Rajaure, S., Bijukchhen, S., Ichiyanagi, M., Dhital, M.R., et al., 2016. Strong ground motion in the Kathmandu valley during the 2015 Gorkha, Nepal, earthquake. Earth Planets Space 68 (1), 10.

United States Geological Survey (USGS), 2015. M7.8. <http://earthquake.usgs.gov/earthquakes/ eventpage/us20002926#general_summary> (accessed 21.06.15.).

Wang, D., Mori, J., 2016. Short-period energy of the 25 April 2015 Mw 7.8 Nepal earthquake determined from backprojection using four arrays in Europe, China, Japan, and Australia. Bull. Seismol. Soc. Am. 106, 259–266.

Wang, F., Miyajima, M., Dahal, R., Timilsina, M., Li, T., Fujiu, M., et al., 2016. Effects of topographic and geological features on building damage caused by 2015.4. 25 Mw7. 8 Gorkha earthquake in Nepal: a preliminary investigation report. Geoenviron. Disasters 3 (1), 7.

Wang, K., Fialko, Y., 2015. Slip model of the 2015 Mw 7.8 Gorkha (Nepal) earthquake from inversions of ALOS-2 and GPS data. Geophys. Res. Lett. 42, 7452–7458.

Wright D, 1877. History of Nepal. Reprint, Calcutta, India: Ranjan Gupta, 1966.

Yagi, Y., Okuwaki, R., 2015. Integrated seismic source model of the 2015 Gorkha, Nepal, earthquake. Geophys. J. Int. 190, 1152–1168.

Zare, M., Kamranzad, F., Lisa, M., Rajaure, S., 2017. A seismological overview of the April 25, 2015 Mw7. 8 Nepal earthquake. Arab. J. Geosci. 10 (5), 108.

Zhang, G., Hetland, E., Shan, X., 2015. Slip in the 2015 Mw 7.9 Gorkha and Mw 7.3 Kodari, Nepal, earthquakes revealed by seismic and geodetic data: delayed slip in the Gorkha and slip deficit between the two earthquakes. Seismol. Res. Lett. 86 (6), 1578–1586.

Zhang, L., Li, J., Liao, W., Wang, Q., 2016. Source rupture process of the 2015 Gorkha, Nepal Mw 7.9 earthquake and its tectonic implications. Geodesy Geodyn. 7 (2), 124–131.

Zhao W, Nelson KD, Project INDEPTH Team 1993. Deep seismic-reflection evidence continental underthrusting beneath Southern Tibet. Nature, 366:557–559.

CHAPTER 3

Seismic Performance of Buildings in Nepal After the Gorkha Earthquake

Humberto Varum[1], Rakesh Dumaru[1], André Furtado[1], André R. Barbosa[2], Dipendra Gautam[3] and Hugo Rodrigues[4]

[1]Faculty of Engineering of the University of Porto (FEUP), Porto, Portugal [2]Oregon State University (OSU), Corvallis, OR, United States [3]University of Molise, Campobasso, Italy [4]RISCO Polytechnic Institute of Leiria, Leiria, Portugal

3.1 INTRODUCTION

On April 25, 2015, at 11:56 a.m. local time, an earthquake of magnitude M_W of 7.8 hit central, eastern, and parts of western Nepal. The mainshock was followed by several strong aftershocks with magnitudes greater than 6.5, the strongest of which was recorded on May 12, 2015 ($M_W = 7.3$). The Kathmandu valley and surrounding areas were most affected by the earthquake. The Kathmandu valley contains mainly substandard to engineered reinforced concrete (RC) buildings, along with brick masonry and adobe residential buildings, and thousands of heritage sites and monuments. In the valley periphery and regions outside the Kathmandu valley, rubble masonry stone buildings are the most common building type. In addition, hundreds of villages in the regions surrounding the valley are settlements characterized by buildings of vernacular construction, all designed without concern to seismic activity.

The earthquake caused extensive damage to both recent and older constructions in the Kathmandu valley, along with 8790 casualties and 22,000 injured. According to the government of Nepal, 498,852 residential buildings collapsed due to the mainshock and aftershock seismic sequences and another 256,697 residential buildings were partly damaged. Although the damage was primarily to non-engineered to pre-engineered RC and brick/stone masonry buildings, closer analysis revealed that a significant amount of damage was sustained by a very large number of buildings, while many other building types performed surprisingly well following the earthquake.

Impacts and Insights of the Gorkha Earthquake. DOI: http://dx.doi.org/10.1016/B978-0-12-812808-4.00003-1

A detailed description of the seismic performance of existing buildings in Nepal following the April 2015 Nepal earthquake can provide valuable insight into the seismic risk and future opportunities for retrofit and mitigation, not only in Nepal but also in other seismic regions that may be subjected to similar strong shaking. In this context, this chapter presents a brief overview of the damage recorded during a field reconnaissance survey carried out after the two main seismic events. The chapter focuses on both recent RC buildings and older substandard constructions. In addition, detailed descriptions of observed damage to urban masonry building stock and rural vernacular constructions are provided. This chapter presents evidence from the field that justifies the observed seismic performance and enables the depiction of damage modes, which could be insightful regarding future efforts to develop earthquake-resistant constructions and strategies to improve the seismic behavior around the world.

3.2 PERFORMANCE OF REINFORCED CONCRETE BUILDINGS

3.2.1 Structural Description and Materials

RC construction increased drastically over the last few decades in the Kathmandu valley and other major urban centers in Nepal to meet the rapidly increasing settlement of the region. RC construction commenced around four decades ago as an alternative to traditional unreinforced masonry (URM) buildings that lack structural integrity and ductility. According to the National Census of 2011, about 10% of building construction in Nepal is RC, with more than 40% of the total RC construction being concentrated in the Kathmandu valley, as shown in Fig. 3.1.

In Nepal, RC buildings constructed before the implementation of design codes are typically characterized by low concrete quality and poor workmanship, often with inadequate column and beam section sizes, insufficient longitudinal reinforcement, large stirrup spacing, and weak beam–column joints. Furthermore, such buildings frequently employ unreinforced solid masonry infill panels for external and internal partition walls. Significantly, the National Building Code (NBC) neglects the influence of infill in structure design, instead focusing on bare frames. Research carried out by several authors (Mehrabi et al., 1996; Varum, 2003; Dolšek and Fajfar, 2008; Chaulagain, 2015) demonstrated that masonry infill considerably increases strength and stiffness if the seismic demand does not exceed the deformation capacity of

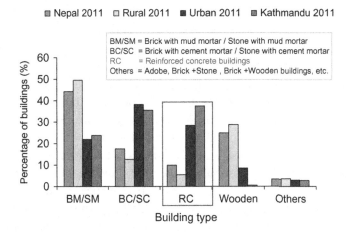

Figure 3.1 Building types in Nepal (Chaulagain et al., 2013).

the structure and decreases the deformation capacity of a structure with respect to its maximum.

Results from site surveys indicate that most RC buildings in Nepal are of 3−5 stories, with column sizes varying from 230 × 230 mm, 230 × 300 mm, to more recently 300 × 300 mm, while beam sizes are approximately (230 × 325) mm, with slab thicknesses ranging 100−150 mm. Most longitudinal reinforcement used in columns of size 230 × 230 mm is 4 ø 12, with 230 × 300 mm columns containing 6 ø 12, and 300 × 300 mm columns (6 ø 16) + (2 ø 12). Primitive buildings are characterized by stirrup spacing of 150 mm on center throughout the column and beam height, whereas more recently constructed buildings present two types of stirrup spacing: 100 mm on center up at the bottom and top one third of the column height at each story and 150 mm on center for the remainder third of the column height. The characteristic strength of the reinforcement is 415 MPa, with concrete strength for earlier buildings ranging from 10 to 15 MPa for structural members. No concrete mix design is typically used for concrete, which therefore allows a great variability in measured concrete strengths.

Even though the RC construction started decades ago, the first design code, known as the *Nepal National Building Code*, was approved only in 2004 and implemented in 2006. Guideline NBC 205 (1994) is useful for regular residential buildings of up to 3 stories, providing ready to use dimensions and details regarding structural members. For other structures, NBC 105 (1994) has been used.

3.2.2 Damage and Failure Modes

As discussed already, earlier building construction in the region did not follow any design code. These structures were built to carry gravity loads and now possess visible structural defects with respect to their seismic design. Although NBC 205 (1994) is applicable to buildings up to 3 stories, site surveys revealed that it is applied to buildings ranging from 3 to 6 stories, with most being irregular 4-story constructions. Structural defects that disturb the load path in the structure detected in the survey, included a short column effect at the staircase landings, lack of continuity of the floor beam due to the presence of the stair landing that decreases the effectiveness of floor diaphragms, and an absence of a seismic gap between adjacent buildings that might result in pounding during ground shaking. Other observed structural defects included issues related to slender buildings that increased expected drift demands, and very different floor heights and floor elevations in adjacent buildings. The latter increased the propensity for these buildings to strike each other, causing the failure of weak stories and potentially even the complete structures.

One of the most common construction issues in Nepal is related to the existence of soft-stories in constructions that are used for residential or commercial purposes. In the case of residential buildings, there is often an opening on the road-facing side used for retail shops that results in lower stiffness and strength with respect to upper stories. Commercial and office buildings often do not have ground infill or even sometimes infill at the basement of the structure, the latter for parking and storage purposes, resulting in considerably lower stiffness and strength at the story at the ground floor, potentially leading to soft-stories during an earthquake, as shown in Fig. 3.2. Poor concrete quality and compaction eventually increases the potential for brittle failure of structural members. The lack of sufficient effective cover in beams and columns may result in the exposure of structural reinforcing steel bars and subsequently lead to corrosion Such a situation may also indicate that the longitudinal reinforcement from the beam is not properly anchored, causing a weak connection at the beam—column joint. The structural deficiencies just mentioned lead to disproportionate effects when these buildings are subjected to moderate to intense earthquakes.

In buildings, each column suffers the same slab deformation during ground motion shaking. However, short columns are stiffer than tall

Figure 3.2 Examples of buildings subjected to soft-story collapse during the Nepal earthquake.

columns, and since the force required to achieve the same deformation is larger for stiffer elements, short columns experience large shear forces. If a short column is not designed for such lateral forces or has insufficient transverse reinforcement, the element suffers significant damage or even collapse during the earthquake shaking, in what is called a *short-column failure*. Such failures were caused by the existence of openings in infill walls (Fig. 3.3A and C) and at mid-story heights in staircase landings (Fig. 3.3B).

Pounding failure observed during the reconnaissance was typically attributed to the lack of a seismic gap between adjacent buildings. Each building has its own natural period of vibration, given approximately by 0.1 N (as per IS 1893:2002)(Indian Standard, 2002) for RC-MRF, where N denotes the number of stories. It is therefore apparent that higher buildings have a longer fundamental period. In addition, the fundamental period of vibration also depends on the stiffness provided by structural and nonstructural elements. The typical building type in the Kathmandu valley is rows of buildings without a sufficient seismic gap, with no uniform interstory height, number of stories, or structural element sizes. Due to these variations, during ground shaking, such structures vibrate at different periods and phases, increasing the chance that they strike each other in an effect known as *pounding*. Under such conditions, the probability of failure of a weak structure is

Figure 3.3 Examples of short-column failure due to the existence of (A) staircase landing and infill, (B) staircase landing beam, and (C) infill panels with openings.

higher. Building failure due to pounding followed by soft-story collapse. Similarly, a significant difference in floor level elevation between two adjacent buildings can also lead to pounding failure.

The so-called beam—column joint failure mechanisms were also observed in Nepal. Beam—column joint failures were attributed to the placing of concrete during two different periods and to insufficient transverse reinforcement and poor reinforcement detailing. The first attributed reason is related to the addition of a further 1—3 stories 5—10 years after the construction of the original building without taking into account the condition of the concrete contact surface, producing weak links with large voids between existing and added concrete. Moreover, for the second attributed reason, the lack of proper lap splicing of reinforcement and insufficient splice lengths explain the beam—column failures, as shown in Fig. 3.4A.

Last, the pancake-type failure was observed in several damaged buildings during the reconnaissance performed following the earthquake. This is often attributed to soft-story mechanisms, but at least in one case this could be explained to the existence of weak beam—column joints.

In older RC building design and construction, it was assumed that the beams should be stronger than the columns. As a result, preengineered RC buildings in Nepal possess a small column section (230 × 230 mm) when compared to the beam size (230 × 325 mm), leading to a weak column-strong beam design. This design approach leads to formation of shear failures or formation of plastic hinges in the columns. This design approach, along with insufficient stirrup

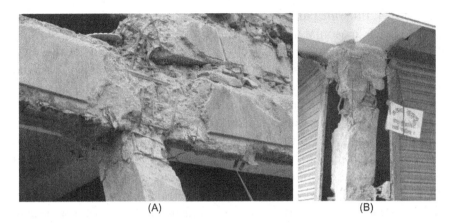

(A)   (B)

Figure 3.4 Failure caused by (A) weak beam—column joint and insufficient reinforcement detailing and (B) poor reinforcing steel bar detailing.

spacing and small stirrup diameter, results in poor confinement of the RC sections and premature reinforcing steel bar buckling (Fig. 3.4B).

Detailing problems recorded during the reconnaissance included insufficient lap-splice lengths that lead to bond-slip failures. For building structures, the modified rule of thumb (MRT) Nepal building code delineates lap-splice detailing requirements, specifying that column lap-splice should be carried out at floor mid-height. In addition, the lap-splice length should be 60Φ (bar diameter), and the number of lap-spliced bars should not be more than 50% of longitudinal reinforcement. However, in Nepal, these guidelines are not enforced on site due to lack of enforcement policies, lack of accountability, or even fear of conflicts arising between contractors and engineers. As a result, the lap-splice of the longitudinal reinforcing steel in the columns and beams are often developed near the beam—column joints with an insufficient lap-splice length. Such sections are the weakest zone of any structural element and failure occurs with concentrations of the deformations at the sections, where bond-slip failures develop. When seismic forces act on these sections, the column will fail, rupturing the concrete and leading to buckling of the longitudinal reinforcing bars. Examples of structures that collapsed during the Gorkha earthquake due to such construction work and poor concrete quality were observed.

Masonry infill panels can increase the seismic capacity of a structure 3—4 times with respect to that of the RC bare frame (Rai, 2005), thus attracting large seismic forces during earthquake loading.

Masonry infill wall panels consist of brick masonry units and mortar joints. Brick masonry units are usually brittle and weaker in tension than in compression, which results in the infill walls being weaker in a biaxial tension—compression stress state than under biaxial compression—compression stress states. The failure of the infill panel is also influenced by the presence of mortar joints. Depending on the orientation of the mortar joints with respect to the applied loading, failure can take place either in the joint only or via a combined mechanism in the mortar and masonry unit. When the stresses are parallel to the bed joints, failure occurs along the interface of brick and mortar joint. Due to these possible modes of failure of the brick and mortar, masonry infill panels are very brittle in nature and in-plane cracking may take place in shear or flexural modes of deformation (Varum et al., 2017). Infill walls also experience both in- and out-of-plane forces simultaneously during an earthquake. Examples of in-plane failures caused by flexural cracking and shear cracking are shown in Fig. 3.5. These failure mechanisms are commonplace in masonry infill panels found in Kathmandu, as the strength of typical masonry units is weaker at the mortar joint in comparison to the surrounding frame. In contrast, out-of-plane failure was not observed as frequently due to the relatively large thickness of infill panels.

At the top of columns, shear failures that were induced by the presence of a masonry infill panel was a common mode of structural failure seen after the Gorkha earthquake. This type of failure was observed mostly near beam—column joints. The damage patterns observed are consistent with those expected for strong-infill and

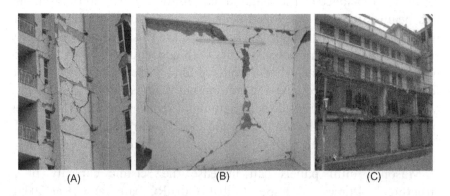

(A)   (B)   (C)

Figure 3.5 Examples of infill masonry failure due to the Gorkha earthquake.

weak-frame modes of failure, including in-plane diagonal shear damage to the infill masonry wall panels and shear failures of RC columns near the beam–column joints. As described before, insufficient shear detailing at column ends was also commonplace, which exacerbates this failure mechanism.

3.3 PERFORMANCE OF UNREINFORCED MASONRY BUILDINGS

3.3.1 Structural Description and Materials

URM buildings constitute the majority of buildings in Kathmandu and surrounding urban areas. These types of structure can be considered nonengineered, as they were built before the existence of modern building codes, with most constructed spontaneously without any support from engineers in their design. URM structures are characterized by the use of poor materials such as solid clay bricks and mud mortar and in a few cases the use of concrete blocks and cement mortar, with material types used varying based on location and building age. For the URM structures, during the reconnaissance survey, the team observed a range of masonry wall thicknesses, varying from 500 to 750 mm, composed of at least a three-wythe wall (three layers of bricks) filled with mud mortar, as shown in Fig. 3.7A. The façades of traditional buildings are typically made from fire clay bricks with a smooth finish, while the inner face is of sundried clay bricks. During the reconnaissance it was also observed that no connection was used between the layers at the time of construction, frequently resulting in the collapse of external walls (Fig. 3.6). Some buildings possess very complex wall systems, with an irregular distribution of bricks and a hard to determine or even undetermined number of layers; due to the weak characteristics of the mortar, the seismic behavior of such buildings is typically very poor.

Floors and roofs are constructed using timber elements, although the former can vary significantly between adjacent and apparently similar buildings. Timber floor joists are common, spanning one or two directions in older buildings, and are built using simple battens or joists upon which timber planks are laid (Fig. 3.7B). These in turn support the final floor finish. Deficient or in some cases no connections between these horizontal structural elements and the masonry walls were frequently observed.

Figure 3.6 URM buildings construction details: (A) masonry wall layer disposition; (B) connection between timber floor and transverse wall; (C) building block.

Figure 3.7 URM earthquake damage and failure modes: (A) out-of-plane collapse of façade wall and insufficient connection between timber floor and masonry walls; (B) out-of-plane collapse of façade wall with bulging of exterior walls; (C) insufficient connection between floors and masonry walls, and out-of-plane collapse of the façade wall of a 3-story building.

URM structures usually comprise 2–5 stories and are commonly constructed in adjacent blocks. This practice contributed to protecting some buildings from collapse, since the seismic response comprised the entire block, compensating for the insufficient capacity of each individual URM building, as shown in the example in Fig. 3.7C. Moreover, in the case of multistory buildings, masonry wall thickness is not uniform throughout the building height but rather decreases from the ground to the top stories, resulting in structural irregularities with elevation. In some cases it was observed that the ground floor was used for commercial purposes, with the reduction in the number of masonry walls in such floors causing structural irregularity and leading to soft-story types of collapse (Gautam et al., 2016).

3.3.2 Damage and Failure Modes

Almost all URM buildings were found to contain no reinforcing bands, such as sills, lintels, or gables, at any level. Due to a lack of proper bonding in the masonry load-bearing walls, out-of-plane collapses were more commonly observed in the Kathmandu valley and other settlements. In most URM buildings, the orthogonal walls were found to show incompatible deformations due to a lack of any proper connection between two perpendicular walls, exhibiting poor integrity. Furthermore, due to an absence of integration between several members within the building structural components, out-of-plane failures were also more intense than any other type of failure in some of these cases, as shown in Fig. 3.7.

Diagonal cracking at the corners of openings and in the center of wall segments, as illustrated in Fig. 3.8, was caused by stress concentration at the corners of windows and doors, as well as the absence of sill and lintel bands. Vertical cracks recorded in the center, ends, and corners of walls reflect insufficient or absent bonds at continuous vertical joints (wall-to-wall connections).

Partial and total collapse of buildings was observed in many buildings in Bhaktapur, where entire neighborhoods were destroyed due to

(A)   (B)   (C)

Figure 3.8 URM earthquake damage and failure modes: (A) diagonal and upper cracking due to vertical stresses; (B) diagonal cracking; (C) diagonal cracking due to stress accumulation associated with openings.

the accumulation of various phenomena, such as in- and out-of-plane failures, corner effects, pounding, torsion, and warping failure caused by building irregularities in plan and elevation, reduced space between two adjacent buildings, imbalance in the sizes and positions of openings in walls, and improperly tied roofing material (Shakya and Kawan, 2016).

Finally, as was the case for RC structures, the use of the ground floor for commercial purposes was also observed in URM buildings, creating vertical irregularity in terms of the existence of openings and the placement of interior walls, and leading to the concentration of earthquake damage in this floor, in what is known as the *soft-story failure mechanism*. In some cases, this phenomenon also resulted in the total collapse of many buildings across the Kathmandu valley region. A combination of factors such as the poor quality of material used in construction, building type, and a lack of structural integrity, resulted in building vulnerability to severe damage and collapse. In addition, walls inadequately anchored to the floor or roof diaphragm exhibited large cracks and in some cases, had collapsed entirely.

3.4 VERNACULAR AND RURAL CONSTRUCTIONS

3.4.1 Structural Description and Materials

Nearly 40% of the total housing stock of Nepal is composed of rubble stone masonry buildings. Such buildings are nonengineered and owner built by untrained local masons. The immediate availability of construction materials is of the highest priority during construction; seismic considerations are generally not considered with the rare exception of a few traditional measures, such as timber bands. Vernacular constructions do exist, although such buildings are outnumbered by those built using traditional dry stone or mud mortar in rubble stone. The field reconnaissance undertaken in central Nepal after the 2015 Gorkha earthquake revealed that wall thicknesses typically vary between 390 and 530 mm, with two-way walls being most common, as shown in Fig. 3.9A. The material found between the brick layers also varied, with stone chips observed in the case of dry stone masonry buildings and segregated mud mortar in the case of mud mortar stone masonry construction. Rural stone masonry constructions in Nepal are predominantly 1−3 story structures, as shown in Fig. 3.9B, with shallow foundations, although a few 4-story buildings can also be found in

(A)	(B)	(C)

Figure 3.9 Typical features of rural buildings: (A) heavy dry stone masonry wall; (B) two-storied building subject to in- and out-of-plane collapse; (C) heavy roofing stones.

the middle mountain region of Nepal. Rural constructions are usually surfaced with mud plasters both internally and externally, however, a countable fraction of such buildings can be found without plastered walls. Although such building practice is economic and does not require skilled manpower, the seismic capacity of the resulting buildings is inherently low and falls under the EMS-98 vulnerability class A. As sites are selected without any engineering considerations, rural buildings are generally constructed on sloping terrain due to a lack of flat areas in the middle to high mountains of Nepal. Timber joists supported by structural walls are designed to support the floor and roof. Roofing material varies with the elevation and the economic status of the building owner, with the middle mountains characterized by corrugated galvanized iron (CGI) sheets or thatched roofing and the high mountains by heavy stone roofing (sliced stones), shown in Fig. 3.9C, or mud blocks. Rural stone masonry buildings are typically standalone rectangular constructions with a roof sloping in two directions; masonry partition walls are rarely provided. Where the latter are present, they are generally the same thickness as the structural walls. The field reconnaissance revealed the use of irregularly shaped and sized stones, with timber and bamboo materials limited to elements such as joists, purlins, and rafters.

3.4.2 Damage and Failure Modes

Out of the 474,025 collapsed rural buildings recorded in central Nepal, approximately 90% were constructed of rubble stone, while nearly 70% of the 173,867 damaged buildings were rural stone masonry constructions. It is therefore interesting to note that this type of building was more susceptible to collapse than to minor or moderate damage after the Gorkha earthquake, a finding similar to that recorded after other events in the region, such as the Bihar-Nepal earthquake of 1934 and the Udaipur earthquake of 1988. Observation of more than 10,000 rural buildings revealed that their orthogonal walls were behaving discrepantly (Fig. 3.10B), as was the case for URM buildings in the Kathmandu valley, with wall collapse predominantly occurring in the out-of-plane direction (Fig. 3.10C) in rural stone masonry buildings. In addition, heavy and untied gable construction also led to gable collapses in many buildings in the middle mountains (Fig. 3.10A). Lack of proper maintenance was one of the major causes of damage in most rural settlements in the middle mountains, leading to severe damage and in some cases collapse. Indeed, buildings aged 100 years or more were found to be still in use, with no periodic strengthening or repairs ever undertaken. Delaminated wall wythes, both internally and externally, revealed that mortar was no longer binding the two units and

(A)   (B)   (C)

Figure 3.10 Damage in rural stone masonry buildings: (A) more than 95% of buildings collapsed in the epicentral Barpak village; (B) separated orthogonal walls; (C) out-of-plane collapse of structural and gable end walls.

that the layers were behaving independently. In the case of dry stone masonry constructions, walls were composed of stacked stones in which the layers were not linked to each other. In summary, the major causes of damage of varying degrees to rural buildings during the Gorkha earthquake included a high weight concentration on walls and roofs, a lack of structural integrity, heavy gable construction, poor mortar quality, a lack of seismic provisions and detailing, stress concentration in corners and openings, topographical and ridgeline effects, sloping foundations, as well as progressive damage due to continued aftershocks.

3.5 OTHER BUILDING TYPES

As stone is not abundantly available in the confluence region of the plains and middle mountains of Nepal, timber buildings are typically constructed in these regions. Indeed, nearly 25% of the total housing stock in Nepal consists of timber buildings, while a minor fraction (approximately 2%–5%) are constructed of wattle and daub or bamboo. During historic strong to major earthquakes, including the great Bihar-Nepal earthquake of 1934, timber, wattle and daub, and bamboo houses have shown excellent seismic performance; a similar situation was also the case after the 2015 Gorkha event, with no cases of damage reported for these types of constructions. Timber houses are typically 1–3 story rectangular constructions, with wooden posts and beams used as part of the structural system. The initial stage of timber building construction is similar to that of the RC skeleton system, with wooden frames fixed in place. However, in timber houses, the openings are then filled with knitted bamboo sheets (known locally as *ikra*) or with wooden planks. Due to their lightweight construction, ductility, and adequate orthogonal connections, the resilience of such structures has been undeterred during each significant earthquake in Nepal. Wattle and daub buildings are primarily single-story rectangular constructions, found mainly in the southern plains and lower mountains of Siwalik. Such construction is preferred as a low-cost housing solution in the region due to the wider availability of bamboo than timber. The construction of wattle and daub houses is similar to that of timber buildings. In this construction type, first, bamboo posts are fixed and knitted *ikra* is placed in the openings between the posts. Thermal insulation is typically achieved using roof tiles, although CGI sheets can also be found. Wattle and daub houses in Nepal have proven similarly resilient against strong major earthquakes, including those of 1934, 1988, and 2015. Both timber and wattle and daub constructions can be classified as belonging to EMS-98 vulnerability class D.

3.6 CONCLUSION

This chapter provides insight into the damage observed during a field reconnaissance trip following the April 25, 2015, Gorkha earthquake in Nepal. RC buildings that were not properly designed to resist the seismic forces and suffered extensive damage, including partial or complete collapse, mostly due to vertical irregularities in their construction that caused stiffness differences and subsequent soft-story mechanisms, as is often associated with nonengineered structures. However, many well-designed tall RC buildings also presented significant nonstructural damage, particularly in masonry infill walls. The extensive damage and large number of URM buildings and vernacular constructions that collapsed can be attributed to poor materials and a lack of construction detailing and construction practices that improve structural behavior of such buildings when subjected to ground shaking. These buildings often exhibited vulnerabilities related to weak or nonexistent connections between the walls and floors or roof, which can lead to improper transmission and distribution of stresses among the various elements. This in turn results in a significant reduction in the strength of the walls when subjected to out-of-plane demands, causing significant damage and collapse.

ACKNOWLEDGMENTS

The first and third author would like to acknowledge the support financially support by Project POCI-01-0145-FEDER-007457—CONSTRUCT—Institute of R&D In Structures and Construction funded by FEDER funds through COMPETE2020— Programa Operacional Competitividade e Internacionalização (POCI)—and by national funds through FCT (Fundação para a Ciência e a Tecnologia), namely through the research project P0CI-01-0145-FEDER-016898—ASPASSI—Safety Evaluation and Retrofitting of Infill Masonry Enclosure Walls for Seismic Demands. In addition, the fourth author would like to acknowledge the support of the United States National Science Foundation Award #1545632 and Kearney Faculty Scholar funds from Oregon State University. The sixth author would like to acknowledge the financial support to the research Unit RISCO (FCT/UID/ECI/04450/2013) also funded by FEDER funds through COMPETE2020—Programa Operacional Competitividade e Internacionalização (POCI) and by national funds through FCT (Fundação para a Ciência e a Tecnologia). The opinions expressed in this chapter are those of the authors and do not necessarily represent those of the sponsors.

REFERENCES

Chaulagain, H., 2015. Seismic Assessment and Retrofitting of Existing Buildings in Nepal. PhD thesis, Civil Department, University of Aveiro.

Chaulagain, H., Rodrigues, H., Jara, J., Spacone, E., Varum, H., 2013. Seismic response of current RC buildings in Nepal: a comparative analysis of different design/construction. Eng. Struct. 49, 284–294.

Dolšek, M., Fajfar, P., 2008. The effect of masonry infills on the seismic response of a four-storey reinforced concrete frame—a deterministic assessment. Eng. Struct. 30, 1991–2001.

Gautam, D., Rodrigues, H., Bhetwal, K.K., Neupane, P., Sanada, Y., 2016. Common structural and construction deficiencies of Nepalese buildings. Innovative Infrastruct. Solutions 1, 1.

Indian Standard, 2002. Criteria for Earthquake Resistant Design Structures (fifth revision), ed. 9, Bahadur Shah ZafarMarg, New Delhi, IS 1893 (Part 1).

Mehrabi, A.B., Benson Shing, P., Schuller, M.P., Noland, J.L., 1996. Experimental evaluation of masonry-infilled RC frames. J. Struct. Eng. 122, 228–237.

NBC 105, 1994. Seismic design of buildings in Nepal. Nepal National Building Code, HMG/Ministry of Housing and Physical Planning, Department of Building, Kathmandu, Nepal.

NBC 205, 1994. Mandatory rules of thumb reinforced concrete buildings without masonry infill. HMG/Ministry of Housing and Physical Planning, Department of Building, Kathmandu, Nepal.

Rai, D.C., 2005. Review of Documents on Seismic Evaluation of Existing Buildings. Department of Civil Engineering, Indian Institute of Technology Kanpur, India.

Shakya, M., Kawan, C.K., 2016. Reconnaissance based damage survey of buildings in Kathmandu valley: an aftermath of 7.8 Mw, 25 April 2015 Gorkha (Nepal) earthquake. Eng. Failure Anal. 59, 161–184, 1.

H. Varum, 2003. Seismic Assessment, Strengthening and Repair of Existing Buildings. PhD thesis, Universidade de Aveiro.

Varum, H., Furtado, A., Rodrigues, H., Oliveira, J., Vila-Pouca, N., Arêde, A., 2017. Seismic performance of the infill masonry walls and ambient vibration tests after the Ghorka 2015, Nepal earthquake. Bull. Earthquake Eng. 15, 1–28.

CHAPTER 4

Response and Rehabilitation of Historic Monuments After the Gorkha Earthquake

Kai Weise[1], Dipendra Gautam[2] and Hugo Rodrigues[3]
[1]ICOMOS Nepal, Kathmandu, Nepal [2]University of Molise, Campobasso, Italy [3]RISCO Polytechnic Institute of Leiria, Leiria, Portugal

4.1 INTRODUCTION

This violent and seemingly chaotic tectonic collision created a landscape that reflected the cosmic order, the abode of the gods. In the foothills of the mighty Himalayas, the lake of Nepal Mandala and the Kathmandu valley was created, which was Naga-vasa-hrada the kingdom of the serpent king or Nagaraja. It was flanked on either side by seven holy rivers of the Koshi watershed to the east and the Gandaki watershed to the west (Weise, 1992). The earthquake that allowed for the waters of the lake to drain out along the Bagmati River possibly for the first time since some 18,000 years ago is interpreted as Manjushree cutting through the hill with her mighty sword. This shows how closely the people understood their environment and geological processes were attributed to the creative energy of the gods (Smith, 1978).

The historic settlements within the Kathmandu valley find their origins in the Licchavi period in the first millennium, but the visible remains are from the Malla period. Though the buildings were rebuilt over the centuries and might not be much earlier than the 17th or 18th century, some of the reused wooden elements could date back to several centuries earlier. These compact Newari settlements were built on higher ground, where possible on stable ridges protruding above the fertile sediment deposits. The agricultural land was given priority and categorized as per the type, number of crops, and overall yield, with the highest ranking given to the wet paddy crops. The settlements were strategically located, built in compact form with clear hierarchies in public spaces and monumental buildings (Nepali, 1965).

Impacts and Insights of the Gorkha Earthquake. DOI: http://dx.doi.org/10.1016/B978-0-12-812808-4.00004-3

The people who settled along the Himalayas over the millennia brought with them their beliefs, their culture, and their skills to build houses using locally available material. The resources were always scarce, but considering the cultural determinants and the way of life shaped by the cultural beliefs, the settlers adapted to their physical setting. The numerous combinations of site conditions, shelter design, available material, craftsmanship, functions, and supernatural beliefs created the diversity of vernacular architecture along the Himalayas. Throughout history it was through recurring tests of endurance and trial that communities learned to improve their cultural expressions and create a resilient cultural environment.

Similarly, in Kathmandu during the early part of the second millennium CE, the traditional buildings were first adapted to fire hazards by introducing a system of brick fire walls that stopped the spread of fires from one building to the next. These brick and timber buildings were then phasewise adapted to withstand earthquakes by inserting wooden ties and pegs to dampen the seismic forces. Innovative solutions were used to ensure structural stability against earthquakes, for example by building square timber temples laced with wooden bands on high-stepped plinths that functioned as base isolations.

The Kathmandu valley has been subject to regular seismic activity, and the communities have been responding to this accordingly. Since the large earthquakes seem to occur only every 80–100 years, there is a generation gap leading to loss of memory of the previous event and thereby loss of knowledge. The last two earthquakes for which we have some records show that, though there is continuity in certain practices linked to earthquake preparedness, each natural event has led to disasters and a relearning of basic principles. This chapter presents some of these lessons that have been learned the hard way.

4.2 THE PREVIOUS EARTHQUAKE: 1934 GREAT NEPAL BIHAR EARTHQUAKE

The most recent great earthquake that had a disastrous impact on the Kathmandu Valley was the Bihar Nepal Earthquake of 1934 of magnitude 8.4. Though the epicenter was some 200 km to the southeast, intensities of up to X were recorded in the valley. We know from the earthquakes recorded over the past centuries that an earthquake with magnitude of more than 8 occurs on average every 80 years. Despite this impediment the valley has flourished. There is much to learn from the experiences gathered by the inhabitants of the valley over the centuries to understand how they coped with and responded to the hazard.

There is little documentation of the recovery process after the Great Nepal Bihar Earthquake of 1934. A handful of books exist, along with interviews of people who experienced the event. A comparative analysis of the situation before and after the disaster is possible from early photographs. The greatest evidence lies in the physical remains, the traces and scars of the earthquake and subsequent recovery activities. A better understanding of the recovery after the 1934 event might provide a measure for the present recovery phase, with which we are struggling.

We understand that the 1934 earthquake caused great destruction to the towns and villages. In the book written by Major General Brahma S. J. B. Rana a year after the earthquake, *Nepalko Mahabhukampa 1990 BS* (Rana, 1934) there is a chart indicating numbers of buildings and monuments that were affected. Within the Kathmandu valley 55,739 houses were recorded to have collapsed or been damaged along with 492 monuments, temples, and shrines. Of these, 56 were in and around Kathmandu, 259 in and around Patan, and 177 in and around Bhaktapur.

Immediately after the 1934 earthquake not many monument and temples were restored to their former grandeur. No comprehensive study seems to have been done on this, however observations provide us with three possible approaches to rehabilitation.

Some monuments were restored to a similar form and design, mainly reusing salvaged materials. This however was not carried out following documentation and minute details of the previous structure. An example would be the 55-window palace in Bhaktapur, which was restored however without projecting the ornate windows on the second floor (Fig. 4.1). This was rectified only in a later restoration in the 1990s.

The second approach was to replace the impressive tiered temples with simple white plastered cubical domed shrines, creating a sanctum for the deity. Many such structures can be seen even today. An example

Figure 4.1 Fifty-five window palace in Bhaktapur after restoration in 1990s.

of such as structure is the Bhai Dega in Patan Durbar Square as well as the Fasu Dega in Bhaktapur (Fig. 4.2). For both these temples plans were being prepared to restore them to their original form, even though only small percentage of the original material is available and the design of the previous structure not fully certain.

The third approach was an entire redesign, such as with Dharara or Bhimsen Tower and Ghantaghar Clock Tower (Fig. 4.3). A large part of the area around Hanuman Dhoka including Juddha Sadak and Sukra Path were areas redeveloped after the earthquake. Even Basantapur Square was cleared of houses, providing a grand view of Hanuman Dhoka Palace from Gaddhi Baithak to the 9-story Basantapur Tower.

Considering the overall result of rehabilitation after the 1934 earthquake, one can say that the result was rather unsatisfactory. This could of course be explained by the limited resources, in the form of both material and expertise. The inappropriate restoration and lack of maintenance carried out over the following decades was the cause of destruction during the recent earthquake. Numbers show a very similar extent of destruction to monuments during the Gorkha earthquake of 2015. Districtwise numbers of destroyed and damaged monuments

Figure 4.2 Fasu Dega in Bhaktapur as rebuilt after the 1934 earthquake.

show Kathmandu with 281, Lalitpur with 129, and Bhaktapur with 72. This gives us a total of 482 monuments, 10 fewer than the ones registered after the 1934 earthquake (Rana, 1934; ERCO, 2015).

4.3 PREPARING FOR THE NEXT EVENTS

Visible and pragmatic efforts for disaster preparedness are needed for Kathmandu Valley World Heritage sites to mitigate earthquake damage in case of future events. Seven monument zones within the Kathmandu valley were inscribed as a single World Heritage site in 1979. These were the three Durbar Squares in Kathmandu, Patan, and Bhaktpur; the two Buddhist Stupa complexes of Swayambhu and Baudhanath; and the two Hindu Temple complexes of Pashupati and Changu Narayan. In 2003, the Kathmandu valley was inscribed on the List of World Heritage in Danger due to uncontrolled development

Figure 4.3 Ghantaghar Clock Tower as rebuilt after the 1934 earthquake.

and loss of historic fabric (UNESCO, 2003). This led to the establishment of an Integrated management plan, adopted by the cabinet of the
Government of Nepal (2007), which allowed for Kathmandu valley to
be taken off the Danger List. The Coordinative Working Committee,
which was established at that time, is still functioning and has allowed
for cooperation between the seven monument zones of the World
Heritage property and continued discussions on disaster preparedness.
Awareness and training programs were carried out; however, more
concrete steps had not yet been implemented.

Since 2012 the Integrated Management Plan has been going
through a participatory process of review and amendment. With the
80th anniversary of the 1934 great Nepal Bihar earthquake, it was
clear that the countdown to the next big earthquake had started.
Several key government officials went to training courses on disaster
risk management. Regular community meetings were held (Fig. 4.4).
Part of the international training course on disaster risk management
for urban heritage sites run by Kyoto-based Ritsumeikan University
was carried out in Kathmandu. In November 2013, a week-long symposium "Revisiting Kathmandu," was organized by ICOMOS
(International Council on Monuments and Sites) Nepal, ICOMOS
Scientific Committee for Risk Preparedness, UNESCO, and the Nepal

Figure 4.4 Training workshop being carried out in Patan.

Department of Archaeology (DOA) with support from the local site managers in preparation to the countdown, linking the discussions between authenticity, management, and community with disaster risk reduction (Weise, 2015a).

One is however never fully prepared for such a formidable display of natural forces. Even though the question of additional strengthening of monuments might be controversial for most conservation experts, the need for maintenance and restoration was clearly witnessed. The system and procedures for immediate response would also have needed to be established. The draft amendment to the Integrated Management Framework document had just been finalized, which included a section on disaster risk management (Government of Nepal, 2015b). The document was sent to the Department of Archaeology less than 5 minutes before the Gorkha earthquake struck.

4.4 THE 2015 GORKHA EARTHQUAKE: IMMEDIATE RESPONSE

On Saturday April 25, 2015, just before noon the 7.8 magnitude earthquake struck. It was an earthquake that seemed to specifically damage vernacular buildings and historical monuments. Villages in 39 districts were affected, with about half a million houses collapsing and a further quarter million being severely damaged (OCHA, 2015). The most badly affected were 11 districts within the area spanning between Gorkha and Dolakha. Listed monuments were affected in 20 districts with 190 being recorded as having collapsed and 663 partially damaged.

The immediate response after the earthquake struck was to look for survivors. There were locations where special events were being held on the Saturday, and when the structures collapsed large numbers were buried. The phenomenon we could observe in most heritage sites in the Kathmandu valley was that people instinctively contributed to salvaging and safeguarding the components of the collapsed and damaged monuments.

The first coordination meeting took place at the UNESCO Kathmandu Office just 5 days after the earthquake, together with the various authorities and stakeholders, as well as organizations involved in the cultural heritage sector. The following week the Earthquake Response Coordination Office (ERCO) was established at the DOA. To ensure that all stakeholders for the preservation of historical monuments were working together with a shared approach, the first two months were declared a response phase. This meant that everything possible needed to be done to prepare the heritage sites for the onslaught of the monsoon. The main building materials such as wood, brick, roofing tiles, and stone along with the artifacts and ornaments that were lying in a pile of rubble needed to be salvaged and stored. Damaged structures needed shoring and protection from the rain. It was decided that a proactive approach would be applied to the World Heritage properties, the sites on the Tentative List and the monuments on the Classified List of the Department of Archaeology. The remaining monuments would need to be left to the communities and local authorities for them to restore, however, providing them with support and expertise where required.

The response phase consisted of initial assessments, salvaging materials from collapsed or partially collapsed structures, and protecting the structures and the remains from rain as well as further aftershocks. This process worked well, with the Department of Archaeology collaborating with the armed forces. There were however complications with some international agencies, which did not seem to understand that

Figure 4.5 Immediate response using heavy equipment, which was inefficient as well as damaged heritage items.

response procedures must be different in historical settings for modern buildings and infrastructure (Fig. 4.5). One such conflict was linked to debris management, where various prominent international organizations helped clear material from historic sites that would have been the material used to reconstruct the monuments and historic buildings.

4.5 INITIAL ASSESSMENT AND PDNA

The initial assessment carried out provided a general overview of collapsed and damaged monuments (ERCO, 2015). The assessment results show that, within the seven monument zones of the Kathmandu Valley World Heritage property, there were 38 collapsed and 157 partially damaged monuments. In addition to these, considering all 20 affected districts, there are 151 collapsed monuments and 474 partially damaged monuments. A separate list was prepared for the Rana period neoclassical palaces, with one collapsed structure and 32 partially damaged buildings. This provides us with overall 190 collapsed monuments and 663 partially affected monuments.

The buildings of the Hanuman Dhoka Palace Museum were in a precarious state, with part of the wing with the King Tribhuvan exhibits having collapsed out onto Basantapur Square. The top three floors of the adjacent 9-story tower had also collapsed. Two thrones, a coffin, and a canon were the largest exhibits located in this section of the museum. We were not in any position to implement provisions requiring elaborate engineering or heavy equipment to transport and put in place large elements. We had to resort to local contractors, local methods, and local security standards. Pipe scaffoldings were erected from which trained soldiers entered and removed the artifacts. The damaged palace structures were propped up using a similar approach of doing the best with what we managed to collect locally.

Swayambhu, the stupa complex on the hillock to the west of Kathmandu, was also extensively affected by the earthquake, with most structures being badly damaged. The main Mahachaitya (the main stupa) had cracks on the hemispheric dome, which needed to be sealed immediately to stop seepage of rainwater. The discussions on how best to accomplish this circled the globe numerous times as ICOMOS experts discussed possible solutions. The result, though very much delayed, allowed for the cracks to be covered in a nonintrusive and reversible manner. The cracks were first filled with acrylic paste then covered with elastic polymer membranes, which needed to be carefully fixed in place to make sure that the many monkeys living around the monument did remove them (Fig. 4.6). In the future, the structure will need to be investigated in detail, using the most advanced technology, to allow for better understanding of the material and construction method of the Mahachaitya structure.

The tantric temple of Shantipur is located just to the north of the Mahachaitya. Only the initiated tantric priest can enter the temple. The structure was recently renovated through an elaborate process of specially initiating a member of the priest's family to help with the work. The structure was damaged by the earthquake; the internal walls had collapsed and the wall painting in the entrance chamber had partially fallen off. The response team had to consider that the inner chambers of the temple could only be viewed by the initiated tantric priests. We managed to collect whatever remained of the fallen pieces of wall painting. The structure itself will require an elaborate procedure of screening off the inner chambers while reconstructing the outer

Figure 4.6 Swayambhu Mahachaitya after waterproofing.

walls (Fig. 4.7). These restrictions are also in place for practically all tantric temples, other examples being the damaged Pratappur and Anantapur shikhara temples to the east of the Swayambhu Mahachaitya.

The Post Disaster Needs Assessment (PDNA) was prepared within this chaotic response period based on early assessments. This was carried out by the various government departments with the support of international agencies spearheaded by the United Nations Development Program and the World Bank. Within 2 weeks a comprehensive document was compiled in preparation for the donor conference on June 25, 2015, exactly two months after the earthquake. This was a grand success, with international pledges reaching $4.4 billion. Great interest was shown to support the culture sector. This seems to have been the outcome of major media coverage of the damaged monuments within the World Heritage property.

The standardized procedures of international donor agencies of preparing the PDNA included for the first time the Culture Sector (Government of Nepal, 2015a). The PDNA provided a rough overview of the incurred damage and loss, with recovery needs being determined at $206 million divided equally over six years. This was 3.1% of the overall recovery need. Whether such statistics can encompass the loss and damage to cultural heritage requires further discourse.

Figure 4.7 Recent work inside the Shantipur temple with the inner sanctum screened off.

Various statements published in the PDNA were neither formulated through extensive consultation nor understood in respect to local significance. "Based on the DRR/BBB principle, the restored historic monuments will have a longer life span in which they will continue to attract the attention of Nepali citizens and foreign visitors alike" (Government of Nepal, 2015a). The Build Back Better principle requires special interpretation for cultural heritage, and linking this to a statement on a longer life span for monuments that have been standing for over a century needs clarification.

Clearly the PDNA was a document that provided a rough overview for the donor conference. The donors seem to understand the circumstances to be purely dependent on financial support. Monuments were listed with assessments as rough as putting figures of $1 million, $2 million, or $500,000 for reconstruction. This has had disastrous consequences: The authorities tendered out these monuments based on the PDNA estimates to contractors with no experience in the restoration of heritage buildings.

4.6 CHALLENGES OF ESTABLISHING A REHABILITATION STRATEGY

An overall rehabilitation strategy was prepared by the Earthquake Response Coordination Unit in close consultation between the Department of Archaeology and ICOMOS Nepal with support from UNESCO's Kathmandu Office (Weise, 2015b). The rehabilitation process however was halted due to the political conflict that arose from the adoption of the new constitution by the government. The protesting parties with support from external powers created a blockade of the southern borders of the country, leading to a lack of even the most basic needs let alone the huge quantities of materials and resources required for reconstruction. Due to political unrest the government was not able to establish the institutional mechanisms required for the reconstruction phase, and the National Reconstruction Authority (NRA) remained in limbo.

The earthquake response in respect to cultural heritage has been strategically segregated into phases. The first phase, of two months, was exclusively reserved for earthquake response, which involved preparing the affected cultural heritage for the oncoming rains. This was followed by the monsoon season, when the rains do not allow much construction work to be carried out. The efforts of the response phase were being monitored especially in respect to the effects of the rains on damaged monuments. The next phase was to focus on planning and research comprising of five approaches.

1. *Legal Approach.* There was an immediate need for the preparation of policies and guidelines. The Post Earthquake Rehabilitation Policy for Cultural Heritage was formulated by a team from the ERCO and was submitted to the ministry for adoption. The Conservation Guidelines for Post 2015 Earthquake Rehabilitation (Conservation Guidelines 2072) were formulated in line with the policy (Government of Nepal, 2016a). The guidelines also look at

sites, monuments, and historical buildings over time and introduce provisions for maintenance and renewal. This was to be augmented with a document defining rehabilitation processes and a related checklist.

2. *Research Approach.* Extensive research is required to better understand the complexity of the sites in historical as well as technical terms. Detailed structural and material research of the damage on the monuments such as the Swayambhu Mahachaitya and Hanuman Dhoka palace should help retain most of the original structure. Geological research was required to study the stability of slopes and soil conditions; however, this has not taken place. Urban archaeology investigated the foundation of collapsed temples and cross-sections of Durbar Squares to better understand the damage in the substructure due to earthquake and the chronology of the sites (Fig. 4.8). Furthermore, the safeguarding and sorting of salvaged artifacts from the Hanuman Dhokha Durbar Square was being carried out in a systematic manner, detailed inventories need to be prepared to understand how much of the materials could be reused. Along with this, the conservation of mural paintings was also carried out.

3. *Planning Approach.* Rehabilitation Master Plans were required for several of the complex cultural heritage sites and historic settlements. These were to be prepared for Hanuman Dhoka, Swayambhu,

Figure 4.8 Archaeological examinations of Kastamandap foundations.

Changu Narayan, as well as Sankhu, Nuwakot, and Gorkha. The Rehabilitation Master Plan would help clarify the multitude of involved donors, managers, supervisors, and the communities. It would also define how and over what time the reconstruction would realistically be carried out. This would require procedures for supporting the restoration of settlements and traditional dwellings.

4. *Practical Approach.* The rehabilitation and reconstruction of the monuments would be possible only if we have knowledgeable and skilled artisans. The master craftspersons must be identified and acknowledged. They must be "living national treasures," as the Japanese do for "keepers of important intangible cultural properties." The system of apprenticeship must immediately be expanded to ensure that sufficient artisans are trained to allow for the restoration of the tangible heritage. This would have to be coordinated with the procurement of appropriate materials. The government must also change the system of tendering and giving such delicate work to the lowest bidder. A system of prequalification, inclusion of skilled artisans, and quality control must be introduced.

5. *Information Approach.* The damage assessment is linked to the collection of a lot of information closely linked to the preparation for postearthquake rehabilitation. This requires a systematic database and easy access to information. For this it was decided to establish a database system using ARCHES as the information platform. The process of establishing the database, working on the adaptation of the software as per local requirements, and the establishment of inventories has been challenging.

The planning process for rehabilitation has begun with an initial time frame of 6 years. The first-year focused on emergency response and building capacity. It was understood that there would not be enough skilled artisans to work on so many restoration projects. This meant that skills training would need to become a central theme of the first-year plan. This would also ensure cultural continuity and resilience. Along with skills, the need for appropriate materials also needed to be considered. This led to a heated discussions on what the approach to reconstruction should be: traditional or contemporary.

Once the approach and principles of restoration were finalized, the guidelines needed to be supported by clearly defined and planned procedures. Each monument needed to be documented and further research might need to be carried out. Traditional rituals and ceremonies needed to be performed during restoration, such as the *Chhema puja* or the asking for forgiveness. The quality of materials as well as craftsmanship was to be guaranteed through stringent supervision and training provided by the local masters.

The reconstruction that took place after the 1934 earthquake shows that in places of lesser importance work was carried out hastily and the workmanship was shoddy (Rana, 1934). Many monuments were never reconstructed, and if a deity needed protection, often a simple white cubical with a dome-shaped roof was constructed, possibly indicating the lack of resources and timber. Plans for reconstruction of monuments in Bhaktapur and Patan were still going on when this latest earthquake struck.

It must be kept in mind that throughout history the built heritage of the Kathmandu valley has undergone regular destruction by the forces of nature. The cycle of destruction and renewal has taken place throughout history and must be accepted as an integral characteristic of the heritage. The value therefore does not lie purely in the material. If the community has the capacity and the will, its cultural heritage will be rebuilt. The constant renewal of the heritage ensures continuity.

The living cultural heritage has persisted and become increasingly more resilient through the lessons learned over time. The close link between the community and the environment has allowed for this system of cyclical renewal to take place. Even though in the cases of many monuments this link has been severed, it is only through the reestablishment of the community as caretaker that any feasible means of safeguarding these monuments can be ensured. If the community can maintain and safeguard the heritage, there is a reason for its continued existence (Fig. 4.9).

4.7 POST DISASTER RECOVERY FRAMEWORK AND ONGOING REHABILITATION

The rehabilitation of the communities and the cultural heritage will take many years. An initial 6-year plan was prepared so that certain targets would be met by July 2021. Though there was a formal system of carrying out the rehabilitation of many of the heritage sites, the informal interventions by the community have been the most critical. The response in

Figure 4.9 The Seto Machhendranath Chariot being pulled through a damaged Kathmandu Durbar Square.

most areas has been controlled, and communities have been obstinate not to give into the dire circumstances. It is this spirit of the communities that will be vital to ensure that recovery will take place rapidly.

The clash between modern engineering interpretations and traditional nonengineered knowledge seems to have come to a head. Reconstruction is being proposed using modern engineering parameters without even properly assessing the performance of the traditional structure or understanding the reason for the damage or collapse. Why did the central timber mast of the Bauddhanath stupa get damaged? Was it because the base of the harmika had been casted using cement concrete? Was the brick masonry in mud mortar in the plinth of Pratappur temple shattered by the recent reconstruction of the superstructure in the more rigid lime-surkhi mortar? Did the upper part of the 9-story palace at Hanuman Dhoka collapse because of the fracturing of a reinforced cement concrete tie-beam introduced in the 1970s restoration? Several tiered temples collapsed that had concrete tie beams. What was the cause of the collapse? Even the collapse of Kasthamandap raises questions concerning earlier interventions rather than design faults after the archaeological investigation.

The lack of understanding of the traditional structures is alarming. In the rush to reconstruct certain monuments, simplified procedures are used. It is important to understand that the restoration project of Kasthamandap in the 1970s covered up the fact that one of the main four central posts was not resting on a saddle stone. At the base of many of the posts the tendons were missing and the holes in the saddle stone filled. The structure probably collapsed because it was not locked to the plinth and was standing on only three out of four main posts (Coningham et al., 2015). We also know that the structure did not collapse immediately, and many people would have survived if they would have moved away. Further research and documentation is required to fully understand what happened.

The NRA was established on December 26, 2015. In preparation for the first anniversary of the Gorkha earthquake, the Post Disaster Recovery Framework (PDRF) was hurriedly prepared (Government of Nepal, 2016b,c). The PDRF had a specific section dedicated to cultural heritage. It was possible to integrate a well-defined strategy and implementation procedures for the rehabilitation of the culture sector, which was official adopted and published by the government of Nepal. Sadly, the document remained in English, to be submitted to donor agencies, but has hardly been read by the local authorities.

4.8 STRUCTURAL DAMAGE ASSESSMENTS AFTER GORKHA EARTHQUAKE

The postearthquake need assessment conducted after the Gorkha earthquake estimated a $169 million loss attributed to damage to tangible heritage sites (Government of Nepal, 2015a). Heritage sites and monument damage was particularly strong in the Kathmandu valley that comprises 7 out of 10 world heritage sites in Nepal. In addition, heritage and monuments outside the Kathmandu valley were also affected at various extents. Nepali heritage sites and monuments are not well discussed and generally underresearched in terms of interdisciplinary approaches that may outline seismic vulnerability precisely. Gautam (2017) highlights four governing factors that affect the seismic performance of heritage and monument structures: ground shaking, periodic maintenance, structural deficiencies, and local site effects. Therefore, interdisciplinary and multicriteria evaluation is needed to understand the seismic behavior and the damage mechanisms, as well as to develop rehabilitation and seismic strengthening measures to be applied specifically to heritage sites and monuments in Nepal. The

Gorkha earthquake highlighted some fundamental seismic vulnerabilities of existing heritage sites and monuments in Nepal. Several reconnaissance missions were carried out after the Gorkha earthquake. A brief overview of damage to heritage sites and monuments can be summarized as follows:

- During Gorkha earthquake, heritage sites and monuments having greater height than other normal heritage sites experience serious damage to collapse. For instance, the 9-story Dharahara tower felled up to the basement (Fig. 4.10A), similarly the 9-story durbar suffered from brittle collapse of some upper stories (Fig. 4.10B). Such damage may be partly attributed to the greater extent of the vertical component as well as low-frequency ground motion occurrence during Gorkha earthquake.
- The neoclassical (Greco-Roman) monuments in the Kathmandu valley were damaged mainly due to lack of ductile components, massive construction, and lack of integrity in the case of orthogonal walls (Fig. 4.10C). In addition to this, the age of such monuments and lack of proper binding between the bricks were leading causes of damage.

Figure 4.10 Damage observations to Kathmandu valley heritage sites and monuments: (A) 9-story Dharahara tower depicting brittle collapse in N−S direction; (B) sandwiched upper stories in 9-story Durbar; (C) damage due to shearing and lack of integrity in Singh Durbar; (D) Kathmandu Durbar Square depicts complete collapse due to lack of adequate connection in substructure−superstructure interface, whereas lightweight and low-height temples remain unaffected. A massive low-height temple shows bifurcated out-of-plane collapse in N−S direction.

- Many pagoda temples, for which walls are designated as the most vulnerable structural components (Shakya et al., 2013), in the Kathmandu valley collapsed due to lack of proper connection between substructure and superstructures and therefore may have experienced discrepant shaking. Such evidence was observed in the Narayan temple in Kathmandu Durbar Square (Fig. 4.10D), where the substructure system (plinth) remained unaffected and the warped timber posts were observed after the collapsed superstructure was removed. The lateral load-resisting systems provided in the Narayan temple would have not been as resilient as that in the Nyatapola temple due to its short anchorage.
- Pagoda temples constructed with less brick masonry (e.g., Pashupatinath) that led to lowered weight survived well during the Gorkha earthquake. In case of Pashupatinath, the rocky strata may have also contributed to its survival during every strong to major earthquake, as noted by Gautam (2017). However, temples constructed with heavy brick masonry walls and having low seismic capacity associated with structural deficiencies were damaged heavily (Fig. 4.10D).
- One aseismic feature identified in the majority of pagodas is the horizontal timber band; however, the efficacy of such a band was observed to depend on its robustness, continuity, and anchorage. Some of the wythes (vertical portions of brickwork) even depicted delamination, even though the timber band was present (Fig. 4.11A).
- Some of the monumental monasteries were found to be constructed with reinforced concrete and brick infills. In such monasteries, damage was attributed to poor ductile detailing, especially in beam—column joints as well as damage due to shearing actions (Fig. 4.11B). Infill walls were not anchored to main structural systems, so the damage should have occurred.
- Several heritage structures in Nepal have not been incorporated under a periodic rehabilitation framework. Even medieval constructions are not rehabilitated for centuries, and the seismic capacity of such structures degrades due to lack of rehabilitation, leading to collapse as in Figs. 4.11C and E.
- The Gorkha durbar is situated in a hammock and lies in a near-field region during Gorkha earthquake. This temple was reconstructed after the 1934 Eastern Nepal earthquake and sustained damages in terms of shear and tensile cracks on a structural wall along with some damage to its roof (Fig. 4.11D). The complex nexus of near-field

Figure 4.11 (A) Delaminated exterior wyth of a temple due to lack of proper binding even though single timber band exists; (B) diagonal shear cracks in infill masonry and heavily damaged beam−column joint due to lack of transverse stirrups; (C) slumped old temple near Bagmati River in the bordering region of Kathmandu and Patan; (D) Gorkha durbar located in a field near the Gorkha earthquake. This durbar is situated in a rocky cliff and sustained damaged in its roof and tensile and shear cracks in masonry walls. (E) Machhindrabahal in Bungamati collapsed in a NE−SW direction as no ductile members were present in entire structure. (F) Krishna temple (Shikhara style) witnessed considerable damage (restricted access) due to main shock although nearby pagodas and brick masonry Shikhara felled.

shaking, topographical amplification, as well as the influence of a rock site deserves further investigation; however, it could be noted that rock site should have minimized the damage extent.

- Shikhara stone temples like the Krishna dewal (Fig. 4.11F) and Chyasing dewal sustained major and minor damages, respectively, whereas a nearby brick masonry Shikhara temple collapsed. Such discrepancy in terms of damage was due to weight reduction and an effective vertical load propagation system in stone temples that is lacking in brick masonry Shikhara. Brick masonry Shikhara were more slender constructions with stacked bricks and the mortar was not holding the brick units properly. Similar observations were made in Fasi dega and Narayan temple in Bhaktapur Durbar Square.
- Dome heritage sites like Swoyambhu and Bouddha witnessed minor damage in terms of shear cracks. Such low damage is attributed to the symmetrical and low rise construction and proper binding. On the contrary, the Ashok stupa in Patan was fissured across an E-W direction that may be due to the soft soil used for dome construction.

• Heritage structures like the 55-window palace, Nyatapola, Bhairav temple in Bhaktapur survived with no or minor damage during the Gorkha earthquake. Periodic rehabilitation of heritage structures minimized the damage in Bhaktapur. During the 1934 earthquake, damage in Bhaktapur was the most severe, and the 2015 Gorkha earthquake altered the scenario at least for major heritage structures.

4.9 IMPLEMENTATION PROCEDURES: THE CHECKLIST

The implementation of many restoration projects has come to a standstill due to the lack of agreed-upon procedures. The standard government procedure, based on the Public Procurement Act, has clearly not been appropriate for monument restoration. There has been a lack of coordination between the various government authorities, particularly the Department of Archaeology, the NRA, and the local governments. Furthermore, there is little communication and support from relevant international agencies in the field of cultural heritage. To facilitate and help coordinate the stepwise implementation, a "Rehabilitation Check-List" has been prepared (Weise, 2016).

The first phase focuses on preparation. The checklist includes documentation, assessment, research, inventory of salvaged materials, and temporary interventions. Only once the preparations are completed to a satisfactory manner, with sufficient documentation, assessments, and research, will work continue to the next phase. The inventories of all salvaged materials from any specific monument or site would have to be prepared and, where possible, their original location determined to allow for reuse of materials where possible and appropriate. Various temporary interventions might still be required even after the initial assessments and response activities; and these will be carried out in an appropriate manner, considering the impact on long-term rehabilitation.

The second phase focuses on design and planning. The checklist includes structural interventions, conservation, material (requirement and supply), artisan (requirement and availability), and implementation planning. The design and planning of interventions consider the structure as well as the ornamentation and focus on both technical and practical considerations. The reuse of salvaged materials goes hand in hand with the required skills in traditional building crafts or use of appropriate modern technology. The required human, material, and financial resources are to be ensured along with the phasewise work schedule, indicating critical paths as well as preparations for the implementation phase.

Once the design and planning has been agreed upon by the respective authorities, Phase 3 implementations begin. The checklist includes rituals, documentation of implementation, supervision and monitoring, handing-over procedures, as well as an audit of quality and finances. This means the procedures must follow traditions while being monitored for compliance to rehabilitation guidelines. The projects will be handed over to the site managers along with an audit on the quality of work as well as finances, which will be made public as soon as it is finalized.

The "Rehabilitation Check-List" will be used as part of any agreement with national and international partners to clarifying the content and schedule of projects. Should a specific party agree to carry out only certain activities on the "Rehabilitation Check-List," the agreement will be finalized only once other partners are determined for the remaining activities. During the implementation process, the same "Rehabilitation Check-List" will be used by the responsible national authorities to monitor the progress and phasewise implementation. Each phase must be completed to a satisfactory degree before the next phase begins. However, there might be certain circumstances where an intervention cannot be designed due to lack of research, requiring the process to return to activities in earlier phases until a satisfactory result is achieved. Phase 3 implementations will not begin without the national authorities agreeing to the design, interventions, and overall implementation planning.

In Nepal, we have various initiatives responding to the damage caused to monuments and historic buildings by the 2015 Gorkha earthquake. Each involved authority, national and international agency, and expert has its own understanding of approach and procedure. There are additionally those who do not seem to have any idea, such as numerous contractors given the task of reconstructing temples. The lack of a mutually agreed-upon and enforced rehabilitation procedure and guideline is causing havoc in the postdisaster recovery of the culture sector. It is high time for the respective authorities to acknowledge these circumstances, reassess the situation, and bring the culture sector rehabilitation back on track.

Figure 4.12 Earthquake damage to Sulamani Temple, Bagan.

4.10 COMPARISON TO THE CHAUK EARTHQUAKE IN BAGAN

Assessments showed that in Bagan 389 monuments were affected by the Chauk earthquake, which struck Myanmar on August 24, 2016 (Fig. 4.12). There were many other monument clusters outside the heritage area of Bagan that were also been affected by the earthquake but these were given less importance. This was even though many were from the Bagan period or even from the earlier Pyu period.

Prioritization in respect to earthquake response within Bagan showed that 36 monuments had major damage and required immediate and comprehensive treatment, while 53 had moderate damage and response was slightly less urgent. The remaining 300 had minor damage, usually to pinnacles and nonstructural elements. A separate assessment of mural paintings and stucco ornamentation lists 94 monuments that were affected (Weise, 2015c).

During the following response phase the monuments underwent rapid assessment while being covered or treated to protect them from rain infiltration. Elements from damaged parts were salvaged to a large degree, even though there is still some work remaining. The most complex task was however to provide structural support (Fig. 4.13). The main problem was to get sufficient experts who could provide instructions and designs for shoring and providing belts to hold

Figure 4.13 Response period covering of buildings with bamboo scaffolding.

damaged structures together. Getting the right kind of materials and equipment was difficult. Some impressive work was carried out on simple structures; however, the large and more complex temples are waiting for more comprehensive intervention.

The response phase was planned for 3 months; however, considering the overwhelming support and progress, there was pressure to move to rehabilitation. Such pressures came from the pagodas with temple trustees who have the support and means to carry out their own rehabilitation projects. They were however waiting to be given instructions on how to carry out such rehabilitation. This acceptance to cooperate with the Department of Archaeology and UNESCO came after statements were made by State Counselor Daw Aung San Suu Kyi to be patient and follow the instructions of the experts.

The rehabilitation phase required an entirely different approach from what was going on as a response to the earthquake. The response phase dealt with short-term immediate measures, which might last for a few years while arrangements are made for more long-term interventions. Rehabilitation itself needed to follow agreed-upon overall conservation approaches that affect the heritage significance of the site. The overall conservation approaches needed to be negotiated with stakeholders.

The planning for the reconstruction phase began with clarifying the overall procedures and providing for different components. The first phase of the process included detailed assessment, documentation, and required research. This information would need to be compiled into a database. The database would then help carry out the second phase, which would be planning the intervention along with detailed designs, material specifications, and quantities. This planning would consider the reuse of salvaged elements along with following agreed-upon rehabilitation guidelines.

Guidelines were adopted for the three categories of monuments, defined as historic monuments, living monuments, and rebuilt structures. In principle, historic monuments are to be conserved, living monuments rehabilitated, and rebuilt structures maintained and, if required, reused. When carrying out the third phase, consisting of the implementation, the respective stakeholders are to get involved. Conservation can be done by experts, but rehabilitation requires an approach with much broader participation of the caretakers, traditional artisans as well as the community.

In Nepal, this process and similar principles have been included in the culture sector PDRF adopted by the NRA (Government of Nepal, 2016b,c) as well as the Post-Disaster Conservation Guidelines adopted by the Department of Archaeology (Government of Nepal, 2016a).

4.11 OVERVIEW OF LESSONS TO BE LEARNED

Discussions on establishing guidelines for the postearthquake response for monuments and historic structures have led to major philosophical differences. There are those who demand that the structures be rebuilt stronger and ensure safety, even if it requires the use of contemporary technology and materials. There are others with the opinion that the introduction of alien elements would go against all traditional principles of construction and destroy the significance of the monuments. A specific example is with the foundations, which are traditionally never reconstructed, since the original rituals determine place and protection. Nevertheless, in some cases these have been removed and replaced with a reinforced concrete mat foundation or with a layer of stone soling.

In respect to authenticity, these reconstructed structures must ensure a credible expression of its original value, which will ensure cultural continuity. At the same time, further research must be done on the traditional materials, technologies, structural systems, and form to devise an improved means of reconstruction that is deeply rooted in the local traditions. The system of construction must ensure that the structure can constantly be renewed, which allows it to persist through time. If any modern technology or materials are used, these components must be reversible, allowing for rectification if considered necessary in the future. It will however be through the motivation and skills of the community that cultural continuity will truly be ensured.

An initial clarification that will be required is consensus on why we are rebuilding the monuments. It could be to safeguard expressions of the local culture and to fulfill the community's requirements for prayer, social gatherings, or festivals. The motivation could be for tourism, for academic reasons, to uphold national pride, or for politicians to show their commitment to the sentiments of the people. If all these motivations could be channeled in a single direction, it would be possible to achieve wonders.

Various important lessons are to be learned from the trials and tribulations in response and rehabilitation after the Gorkha earthquake:

1. *Responsibilities and coordination*: The overlapping of authority often led to friction and inefficient use of limited resources. Any ambiguity will lead to chaotic circumstances with work being duplicated or goals contradicting each other. A clear line of command, clarification of authority, and coordination of all related authorities must be established where possible before the disaster and, if not, then as soon as possible.
2. *Response procedures for salvaging and storing*: There was great loss of original material due to debris management efforts that led to most materials from a collapsed structure being removed from the site. The procedures for identifying reusable material and salvaging important elements must be institutionalized with all authorities being made aware.

3. *Inventory and prioritization linked to vulnerability assessments*: The initial response was delayed and confused due to lack of prioritization and listing of monuments. Detailed inventories must be prepared before the disaster with guidelines on prioritization of monuments for immediate response. Ideally, this would be linked to vulnerability assessments so that certain scenarios can already be prepared before the disaster.

4. *Documentation and database*: The lack of documentation and means of storing all the information collected during the various phases of assessments has been a major obstacle in the rehabilitation phase. A comprehensive database with detailed documentation of the monuments would be essential in preparation for any disasters. The database would also be the basis for planning response while collecting information on assessments. This would also be the basis for rehabilitation planning. This would go hand in hand with research in archaeology, structural engineering, conservation, materials, geology, artisan skills, and ornamentation.

5. *Understanding of the monuments*: The lack of understanding of traditional construction has led to inappropriate response and unnecessary loss of structures and material. Discussions on responding to monuments that are damaged as well as finding appropriate means to rehabilitation requires a detailed understanding of the monuments in respect to their structure, materials, and craftsmanship. Such studies must be carried out in preparation for a disaster and information must be made accessible to the response teams.

6. *Mitigation measures linked to maintenance and restoration*: The reason for damage or collapse of most monuments was lack of maintenance and inappropriate interventions in the past. Disaster mitigation measures are not always easy to demarcate for specific sites; however, basic procedures must be linked to regular maintenance and appropriate restoration. Interventions on monuments must be carried out, keeping in line with the highest of standards and quality. Several paradigms like the 55-window palace and Nyatapola highlight that periodic rehabilitation matters a lot in terms of protection of heritage structures. The seismic capacity of heritage structures can be enhanced considerably with traditional measures being compliant with patrimonial issues; however, implementation of such measures needs to be justified through interdisciplinary and multiapproach analysis studies and these should include experimental tests as well as numerical modeling.

7. *Planning response and rehabilitation*: The response phase was never fully completed while unplanned rehabilitation was forced to begin without adequate preparations. The strategic planning for a time-bound response period focusing on immediate requirements is required, which a fixed date to move onto a rehabilitation phase with mid- to long-term goals.

4.12 CONCLUSION

The observations and approaches after the 2015 Gorkha earthquake depict that the seismic vulnerability of monuments and heritage sites in Nepal is very high; therefore, proper actions in terms of preparedness are needed immediately. Damage in case of heritage sites highlight the severe state even in case of low ground shaking; hence, future events with greater shaking may lead to collapse of most of the heritage sites and monuments. Systematic review of the actions before and after the earthquake is presented in this contribution and some insights from Bagan earthquake (Myanmar) are presented for comparison.

REFERENCES

Coningham, R.A.E., et al., 2015. Post-disaster urban archaeological investigation, evaluation and interpretation in the Kathmandu Valley World Heritage property. Report and Recommendations of a Mission Conducted between 5/10/2015 and 22/11/2015. Unpublished report, Department of Archaeology, Durham University.

ERCO, 2015. Preliminary List of Collapsed and Partially Damaged Monuments, Updated on 30 July 2015. Earthquake Response Coordination Office (ERCO), ICOMOS Nepal and Department of Archaeology, unpublished.

Gautam, D., 2017. Seismic performance of World Heritage sites in Kathmandu valley during Gorkha seismic sequence of April–May 2015. J. Perform. Constr. Fac. Available from: http://dx. doi.org/10.1061/(ASCE)CF.1943-5509.0001040.

Government of Nepal, 2007. The Integrated Management Framework for the Kathmandu Valley World Heritage Property. [Published in Nepali] UNESCO, Kathmandu.

Government of Nepal, 2015a. Nepal Earthquake 2015—Post Disaster Needs Assessment, Volume A—Key Findings. National Planning Commission, Kathmandu.

Government of Nepal, 2015b. Draft amendment to the integrated management framework document for the Kathmandu Valley World Heritage property. Unpublished report.

Government of Nepal, 2016a. Post-Disaster Rehabilitation Guidelines. Department of Archaeology, Kathmandu.

Government of Nepal, 2016b. Nepal Earthquake 2015—Post Disaster Recover Framework. National Reconstruction Authority, Kathmandu.

Government of Nepal, 2016c. Nepal Earthquake 2015—Sector Plans and Financial Projections. National Reconstruction Authority, Kathmandu.

Nepali, G.S., 1965. The Newars, an Ethno-Sociological Study of a Himalayan Community. Unit Asia Publications, Bombay, India.

OCHA, 2015. Nepal: Earthquake 2015 Situation Report No. 20 (as of 3 June 2015), e-published on Reliefweb.

Rana, B.S.J.B., 1934. The Great Earthquake of Nepal 1934 [in Nepali]. Jorganesh Press, Kathmandu, Nepal.

Shakya, M., Varum, H., Vicente, R., Costa, A., 2013. Seismic sensitivity analysis of the common structural components of Nepalese Pagoda temples. Bull. Earthquake Eng. 12 (4), 1679–1703.

Smith, W.W., 1978. Mythological History of the Nepal Valley from Svayambhu Purana. Avalok Publishers.

UNESCO, 2003. *Decision of the 27th Session of the World Heritage Committee*, UNESCO Document WHC-03/27.COM/7B. UNESCO, Paris, France.

Weise, K.U.P., 1992. Kathmandu, Metamorphosis and Present Issues. Unpublished thesis, ETH Zurich, Switzerland.

Weise, K.U.P. (Ed.), 2015a. Revisiting Kathmandu, Document on Safeguarding Living Urban Heritage. UNESCO Office in Kathmandu and World Heritage Institute for Training and Research for Asia and the Pacific, Shanghai, China.

Weise, K.U.P., 2015b. Nepal Post-Earthquake Rehabilitation of Cultural Heritage, Phase 2: Planning and Research. Unpublished report, UNESCO Kathmandu, Nepal.

Weise, K.U.P., 2015c. Response to Chauk Earthquake of 24 August 2016, Notes on Mission to Bagan 26 September to 10 October 2016. Unpublished report, UNESCO, Bangkok/Yangon.

Weise, K.U.P., 2016. Rehabilitation Procedures 2072. Unpublished document, ICOMOS Nepal.

FURTHER READING

Government of Nepal, 1979. Nomination Document for Kathmandu Valley. Nomination document, Government of Nepal, Unpublished report.

Sharma, D.R., Shrestha, T.B., 2007. Guthi: community-based conservation in the Kathmandu Valley. UNESCO Kathmandu Office, Unpublished.

CHAPTER 5

Risk Management, Response, Relief, Recovery, Reconstruction, and Future Disaster Risk Reduction

Amod M. Dixit[1], Surya N. Shrestha[1], Ramesh Guragain[1], Bishnu H. Pandey[2], Khadga S. Oli[1], Sujan R. Adhikari[1], Surya P. Acharya[1], Ganesh K. Jimee[1], Bijay K. Upadhyaya[1], Surya B. Sangachhe[1], Nisha Shrestha[1], Suman Pradhan[1], Ranjan Dhungel[1], Pramod Khatiwada[1], Ayush Baskota[1], Achyut Poudel[1], Maritess Tandingan[1], Niva U. Mathema[1], Bhuwaneshwori Parajuli[1], Gopi K. Basyal[1], Suresh Chaudhary[1], Govinda R. Bhatta[1] and Narayan Marasini[1]

[1]National Society for Earthquake Technology-Nepal (NSET), Lalitpur, Nepal [2]British Columbia Institute of Technology, Burnaby, BC, Canada

5.1 INTRODUCTION

5.1.1 General

Nepal faces a multitude of natural hazards. The corresponding risk in terms of human casualty and loss of houses and other infrastructure is extremely high because of the physical and social vulnerabilities that already exist and due to the continued buildup of vulnerabilities, especially to earthquake hazards, as well as due to the lack of any systematic hazard preparedness. Statistics for the past 45 years show that the small-scale, everyday, "extensive" hazards inflict an annual average toll of 618 lives (casualty of about 2 persons per day) and destroy 6133 houses, while the "intensive" risk disasters, the larger ones, kill 145 people annually, and inflict a loss of 7463 dwelling houses annually (NSET, 2016). The 50-year statistics on earthquake disasters cannot represent the actual level of hazards; the history of Nepal, especially that of the Kathmandu valley is strewn with devastating events, at least one per century, that have killed people, from kings to the commoners, and caused tremendous economic and political setbacks. The risk due to natural hazards, especially earthquakes, has been growing especially in the past four decades because of increasing population densities in urban and urbanizing centers, continued weakness of the

Impacts and Insights of the Gorkha Earthquake. DOI: http://dx.doi.org/10.1016/B978-0-12-812808-4.00005-5

national economy resulting in weakness in the management of disaster risk governance, and the lack of awareness and scientific knowledge resulting in low levels of disaster risk perception amid conflicting priorities to meet the basics needs.

Disaster risk management was not an active menu item in Nepal until the country was shocked seriously by the 1988 M6.6 Udaypur earthquake and by the 1993 floods in the south-central part of its territory. The need to develop organized approaches to manage natural hazards was recognized due to the strong impact of these two events and the influence of the International Decade of Natural Disaster Reduction (IDNDR; 1990–1999) and one of its undertakings, the World Seismic Safety Initiative (WSSI). The government of Nepal started several initiatives, including the first National Conference on Disaster Management in 1993, resulting in the first national plan for disaster management, development of a National Building Code (NBC) (1992–94); establishment of a National IDNDR Committee for disaster management, drawing in representation from pertinent government agencies, academia, professional organizations and NGOs; and development and enaction of a National Action Plan on Disaster Management (NAP). Academic institutions and professional organizations also started several initiatives of far-reaching consequences: The Nepal Geological Society (NGS) started officially observing the International Day for Natural Disaster Reduction (IDNDR Day). The National Society of Earthquake Technology—Nepal (NSET) was established as a need was felt to help people and the public sector to reduce the impact of earthquakes.

NSET arrested the attention of people and the government when it published the earthquake damage scenario of Kathmandu valley (NSET, 1999a,b). Later, a study conducted by the Bureau of Crisis Prevention and Recovery of UNDP revealed that Nepal stands at 11th position with respect to its relative vulnerability to earthquakes (BCPR/UNDP, 2004). Earlier, another study on relative earthquake risk in 21 cities located in high-seismicity regions revealed that the Kathmandu valley happened to be the most at-risk city in the world in terms of human casualty due to earthquakes (GESI, 2001).

Nepal participated actively in all processes of the International Decade for Natural Disaster Reduction (IDNDR) and shared its National Action Plan for Disaster Management (MOHA, 1996) in the UN International Conference on Disaster Management in Yokohama in 1994. This opened the way for developing organized approaches for disaster management in Nepal, which was further refined after Nepal became signatory to the implementation of the Hyogo Framework for Action (HFA; 2005–15). Nepal is an active participant in regional and international initiatives in disaster risk management, such as the biannual Asian Ministerial Conference on Disaster Risk Reduction (AMCDRR), the Global Platform for Disaster Risk Reduction (GPDRR), and other forums. As an active member of the SAARC Disaster Management Centre (SDMC) of the South Asian Association for Regional Cooperation (SAARC), Nepal has been promoting the sharing of knowledge and joint initiatives as per the Earthquake Risk Management Road Map of SAARC (SAARC, 2009).

Despite the international attention and support, despite the existence of traditional wisdom in aspects of earthquake risk management, and despite the successes of several initiatives on earthquake risk management, Nepal continues to face ever-increasing risk from earthquakes. This calls for more concerted efforts toward earthquake risk management and institutionalization of the process. Fortunately, earthquake awareness as well as the capacity for earthquake risk assessment and reduction is growing steadily due to efforts of government, nongovernment, academic, and private sector businesses and the support received from international organizations, the United Nations (UN), system and bilateral agencies. This chapter envisions providing information on Nepal's experience in aspects of earthquake risk management.

The following sections provide a brief description on the methods of earthquake vulnerability and risk assessment employed in Nepal, examples of the use of risk assessment to reduce earthquake risk in different sectors, and discusses the achievements as well as the challenges of earthquake risk reduction in Nepal.

5.1.2 Understanding Seismic Hazard of Nepal

Scientific observation of the geology and geomorphology of the Nepal Himalayas has been done since the advent of mountaineering in the first decades of 20th century, including systematic geological research started by European geologists in the late 1930s and 1940s (Heim and Gansser, 1939; Gansser, 1981; Hagen, 1969; Bordet et al., 1972), followed by more detailed studies in the second half of the 20th century (Le Fort, 1975, 1996; Gansser, 1981; Fuchs and Sinha, 1988; Nakata, 1982, 1989; Schelling and Arita, 1991; Liu and Einsele, 1994; Srivastava and Mitra, 1994; Seeber and Armbruster, 1981; Seeber et al., 1981).

J.B. Auden was the first geologist to talk about the earthquake risk of Nepal when he undertook an extensive survey of the effect of the 1934 Bihar-Nepal earthquake (Dunn et al., 1939). Maj. General Braham Samsher Jung Bahadur Rana was another Nepali administrator with scientific acumen who gave a detailed accounts of casualty and damage due to the 1934 earthquake in Kathmandu and the rest of Nepal and, for the first time, indicated the high seismic risk in Nepal and suggested ways to minimize the risk, including the need and method of constructing earthquake-resistant residences (Rana, 1935).

Although some of the early researchers made reference to seismic hazard of Nepal and the social memory of the "scary" 1934 M8.3 Bihar-Nepal earthquake still continued at a household level, there was little organized awareness about seismic hazard in the country. Isolated earthquake episodes, for example that of M7.1 Rasuwa and 6.5 Bajhang earthquakes were treated as emergency response problems and not as any problem threatening the national economy. A great stride in this background was done by the Department of Mines and Geology (DMG), which established the first four short-period seismographs in the periphery of the Kathmandu valley in 1978 and established the National Seismological Centre (NSC) with the responsibility of systematic seismic monitoring. NSC has been conducting seismic monitoring using a nationwide network of short-period seismographs, accelerometers, and more recently GPS instruments and seismic hazard assessment by active fault trenching, and other scientific studies (Sapkota et al., 2013; Pandey et al., 1995, 2002; Adhikari et al., 2015, 2016; Rajaure et al., 2016).

Fig. 5.1 shows the distribution of Peak Horizontal Accelerations that have a 10% probability of being exceeded over 50 years, prepared by the Seismic Hazard Assessment and Risk Evaluation group conducted under the Nepal Building Code Development Project (UNDP/UNCHS, 1994). This assessment prepared a consolidated earthquake catalog for Nepal, inventoried the reported active faults, and analyzed their potential to generate earthquakes, and based on the analysis of the identified point and linear sources, developed the probabilistic

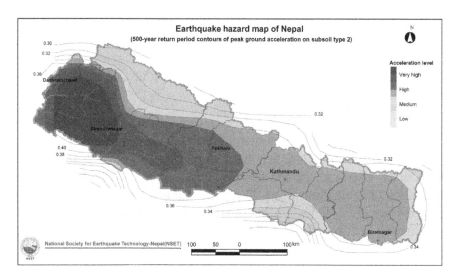

Figure 5.1 Probabilistic seismic hazard map of Nepal showing PGA values with a 10% probability of exceedance in 50 years (BCDP, 1994, recolored and redrawn by NSET).

seismic hazard maps that served as the basis for the formulation of seismic zoning maps and ultimately the NBC of Nepal. For the first time, the country learned about the high earthquake risk and realized the need for conjunctive efforts toward earthquake risk management.

Another milestone in understanding seismic hazard of the Nepal was the seismic hazard assessment done by Department of Mines and Geology, National Seismological Centre, Nepal (DMG, 2002).

5.1.3 Understanding Earthquake Disaster Risk in Nepal

The national discourse on earthquake risk in Nepal owes its origin mainly to two initiatives: the national building code development program in 1992–94, including its components of seismic hazard and risk assessment, study on the building materials and technologies and formulation of the NBC and implementation strategy; and the Kathmandu Valley Earthquake Risk Management Program implemented by NSET in close collaboration with GeoHazards International (GHI) and the Asian Disaster Preparedness Center (ADPC) in partnership with the U.S. Office of Foreign Disaster Assistance (OFDA).

5.1.3.1 Main Sources of Earthquake Risk in Urban Areas in Nepal

The major source of earthquake risk in terms of human casualty in urban areas of Nepal is from possible collapse of buildings as reported by GESI (2001). More than two-thirds of earthquake risk comes from poorly constructed buildings without earthquake resistance. Other two areas that demand attention are enhancement of capacities in emergency medical response and establishment of robust nationwide emergency response system (GESI, 2001).

The 2015 Gorkha earthquake sequence confirmed that buildings are the most lethal hazards in times of earthquakes; a detailed damage assessment of more than 200,000 buildings, conducted by NSET in the immediate aftermath of the Gorkha earthquake, revealed that more than 95% of the people killed during earthquake were inside buildings. This evidence clearly shows the importance of intervention for improving seismic performance of buildings, existing and new.

5.2 DECADES OF EARTHQUAKE RISK MANAGEMENT EFFORTS IN NEPAL

5.2.1 Introduction

The 1988 M6.6 Udaypur earthquake triggered a significant concern in Nepal about the lack of earthquake preparedness and resulted in start of the effort for developing an earthquake-centric building code in 1992. In the course of code development, it was decided to conduct a national level assessment of seismic hazard and risk that provided the scientific basis for the NBC. In addition to input to the building code development, the assessment generated risk awareness among professional communities and the public that resulted in a growing demand for seismic safety. Development and publication of the national building code marks the major thrust to forward the issue of earthquake risk as one of important agendas to be addressed. Under the initiatives of professionals associated with building code development, the NSET, a professional organization dedicated to earthquake risk management in Nepal, was established, having a mission to help people and the government to develop and implement organized approaches for earthquake risk management with strategic objectives of learning modern technologies globally and adapting them to present-day needs in the local context. From the late 1990s, the country witnessed several efforts in earthquake risk reduction and preparedness that were tested in the 2015 Gorkha earthquake. In this section, a few selective major efforts of last two decades in earthquake risk mitigation and emergency preparedness that shaped the country's readiness to respond the 2015 earthquake are discussed.

5.2.2 Kathmandu Valley Earthquake Risk Management

In 1997, NSET started implementation of the Kathmandu Valley Earthquake Risk Management Program (KVERMP), which was instrumental not only in contextualizing international good practices into the local conditions but also in developing confidence among the stakeholders on the possibility of earthquake risk reduction even in the weak economy of Nepal. KVERMP included activities aimed at beginning a self-sustaining earthquake risk management program for Kathmandu Valley. Project components included the following: (1) development of an earthquake scenario and an action plan for earthquake risk management in the Kathmandu valley, (2) a school earthquake safety program, and (3) raising awareness and strengthening institutions. The project was implemented with strong participation by national government agencies, municipal governments, professional societies, academic institutions, and international agencies present in Kathmandu valley through advisory committees and several workshops, seminars, and interviews. The program generated public awareness and policy discourse beyond professional groups on earthquake risk management for the first time in the country.

A simple loss scenario for a possible repeat of the 1934 earthquake shaking level in Kathmandu valley, developed under the program, estimated about 60% of all buildings in Kathmandu valley likely to be damaged heavily, many beyond repair. Simply applying the percentage of the population killed or injured in more recent earthquake casualty figures from cities comparable to Kathmandu valley resulted in an estimate of 40,000 deaths and 95,000 injuries in Kathmandu valley's next major quake with shaking intensities of MMI IX or above (Dixit et al., 2000).

The process and result of this simple seismic risk assessment of capital city provided a grounds for an action plan to deal with the vulnerability that was ever increasing. Kathmandu Valley Earthquake Risk Management Action Plan, an action plan under KVERMP, was created through a wide consultation process involving major stakeholders, including almost all concerned ministries, departments, and critical facilities operators. The purpose of this plan was to assist the government to reduce the earthquake risk by coordinating and focusing risk management activities, many of them to be initiated by NSET at the initial demonstration phase. Specific objectives under the plan's focus were to improve emergency response planning and capability; to improve awareness of issues relating to earthquake risk; to integrate seismic resistance into the process of new construction and improve the seismic capacity of existing ones, including schools and other critical infrastructures, utilities, and the transportation system; to increase experts' knowledge of the earthquake phenomenon, vulnerability, consequences, and mitigation techniques and to be prepared for long-term community recovery following damaging earthquakes.

The School Earthquake Safety Program (SESP) and institutionalizing public awareness for earthquake risk management were other major components of KVERMP. A comprehensive study of the seismic risk of schools and demonstration of an effective retrofit solution for vulnerable school buildings, development and showcase of a community-based framework of comprehensive school safety were major outcomes of SESP that affected the course of disaster risk reduction (DRR) in the education sector. It established the ownership of the concept of seismic retrofitting of school buildings in Nepal and the region. It inculcated the engagement and ownership of the concept by demonstrating technical, economic, social, political, and cultural feasibilities of seismic retrofitting of Nepalese buildings. Two decades later, today, SESP is one of the DRR programs that has conclusively demonstrated its technical, economic, social, and political feasibilities and has become the most attractive programs for funding by the government, small or large funding agencies, and international development partners (NSET-KVERMP, 2010). It has become a national program and the retrofitted school buildings have demonstrated the wisdom of retrofitting as none of the retrofitted school buildings was damaged by the Gorkha earthquake.

The KVERMP program institutionalized public awareness and demonstrated the approach and method of capacity building for seismic risk reduction. It formalized annual Earthquake Safety Day (ESD) through government declaration. The ESD annually organizes comprehensive program targeted to all stakeholders from the school children to the professionals, the private sector to the academia, and from the international development partners to the local government. The ownership of ESD has widened significantly and the activities usually spread all over Nepal for a period of more than three weeks in the month of January every year. The Mason Training (MT), which has now become essential component in program and projects related to safe building construction, was started in KVERMP responding to need to address the nonengineered buildings.

5.2.2.1 Influence and Follow-ups of the KVERMP Program

The KVERMP program influenced directly or indirectly the development efforts in Nepal. It started wider discourse on seismic hazard and risk, attracting research on the vulnerability of the Kathmandu valley and Nepal as a whole. The lineup of priority and "basket" activities in the KV Action Plan influenced the development partners of Nepal to incorporate DRR into the respective development assistance strategies. The KVERMP was able to develop a sense of confidence among the people and the government on the possibility of earthquake risk reduction in a weak economy and fatalistic society. The paradigm shifts from a firefighting mentality of emergency response to that of planning, implementing, monitoring, evaluating, and lesson learning from earthquake risk management activities and programs is a major impact of the program in the direction of sustainable earthquake risk management.

The partnership developed in KVERMP grew continuously even after the program formally ended in 2001. More and more institutions became interested in the DRR initiatives that NSET implemented through projects and programs, which were initially concentrated in Kathmandu valley and later proliferated gradually to other urban and rural settlements outside the valley. During these two decades of earthquake risk management, several milestone programs were implemented that either originated in KVERMP or logically derived from its lessons and outputs. While programs like SESP, annual ESD, and mason training for earthquake-resistant construction have been continued, other major programs, like a program for enhancement of emergency response (PEER), a community-based disaster risk management program (CBDRM), a municipality earthquake safety program (MERMP), a public-private partnership for earthquake risk management (3PERM), a building code implementation program in Nepal (BCIPN), were initiated and successfully implemented. Fig. 5.2 provides a glimpse of output achieved from these efforts in last two decades. It may be noted that growth in implementation accelerated; e.g., the figures in the stigma denote the outcome in the year 2016 while the figures in the petals indicate the cumulative outputs in the past two decades. The usefulness and impacts of these efforts were tested by the 7.8 M_w Gorkha earthquake

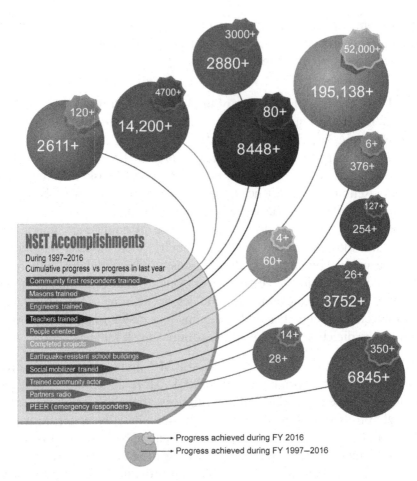

Figure 5.2 Outputs of earthquake risk management efforts made in Nepal in the past two decades.

of 2015. The earthquake showed the efficacy of the approaches, philosophy, and methodologies adopted in carrying out these programs.

The following sections describe three of the initiatives that provided significant support to the people and government of Nepal in the risk mitigation as well as response and recovery and reconstruction phases of the earthquake aftermath.

5.2.3 School Earthquake Safety Program
5.2.3.1 Program Components
The SESP was originally a part of the larger KVERMP program, which aimed at quantifying the earthquake risk in the education sector

and initiating actions for risk reduction. SESP was started in 1998 with an assessment of 623 public schools in the Kathmandu valley and the first school seismic retrofit. While improvement in seismic performance of nonengineered, 2–3 story school buildings of brick construction as a demonstration through seismic retrofit was the aim of the program initially, to highlight the importance of school buildings and to convince the public and the government of the need for compliance to the NBC, the concept of SESP has since evolved wider to additionally include the following:

1. Seismic retrofitting and or earthquake-resistant reconstruction of the main structures and nonstructural elements.
2. Training of masons, carpenters, contractors, supervisors, engineering technicians, and building design engineers and architects on earthquake-resistant design and code-compliance construction technology and seismic retrofitting.
3. Earthquake awareness programs targeting students and community members and teacher training.
4. School disaster emergency preparedness and response planning and exercises of safe evacuation drill and first aid.

5.2.3.2 Program Approach
The SESP adopted a community-based comprehensive safety approach for schools and surrounding communities. The principles of the program are

1. To prefer use of local resources, local masons and contractors, and emphasis on local building types.
2. To foster collaboration among government agencies, civil society, the business community, and local communities by involving them in the decision-making process at all stages.
3. To emphasize technical and financial transparency to the involved stakeholders.
4. To establish a sustainable system so that DRR becomes part of the school system, the community, and the government. This includes establishment of annual vulnerability assessment, regular drill at schools, conducting a program of earthquake awareness to promote community demand for retrofitting and school retrofitting as a model demonstration to convince other public and private schools to follow suit.

5.2.3.3 Institutionalization of SESP

The NSET continued the SESP program after KVERMP with retrofits in an average of three schools and components in some more schools within and outside Kathmandu, aiming to institutionalize the program. Efforts were made to involve the government and gradually build their ownership in the program. In 2010, the Ministry of Education of the government of Nepal incorporated a school building seismic retrofitting program in the annual national plan and program with a budget for seismic retrofitting of 15 school buildings starting from the fiscal year 2011 to 2012. This milestone decision of the government immediately received support from the development partners and international financial institutions. The Nepal Risk Reduction Consortium (NRRC), which draws in membership from various international development partners based in Nepal, included SESP as one of the five flagship programs with the Asian Development Bank serving as the coordinator-lead of the Flagship 1: School. The number of school buildings retrofitted annually grew gradually from 3 per year by NSET to more than 150 per year after 2010 (ADB/GON, 2010). Subsequently, a national strategy for school earthquake safety was developed and enacted. School Safety has become "business as usual" for the Ministry of Education. Currently the school building retrofitting is running in 23 districts outside the Kathmandu valley. Training of teachers on aspects of school safety against disaster has become regular. A process for revising the school teaching curricula to incorporate DRR and preparedness studies has started. A unit to deal with issues of Disaster Risk Management has been established in the Department of Education of the Nepal government. About 250 school buildings have been retrofitted since 2010 as a government-led program. Also 10,000 teachers, and 2000 masons have been trained on aspects of school retrofitting and disaster preparedness; and around 500,000 students have been sensitized to the aspects of disaster safety. The SESP has thus been firmly institutionalized in Nepal.

5.2.3.4 Performance of Retrofitted Buildings in the April 2015 Gorkha Earthquake

While NSET's program of seismic retrofitting of the nonengineered, largely mud-based masonry school buildings of Nepal continued for the past two decades, the methodology adopted was simple and cost effective, utilizing local material and resources. However, little research went into scientific confirmation of the retrofit technology. The

Gorkha earthquake of 2015 provided the much-awaited opportunity to look into the behavior of the retrofitted school buildings to seismic loading.

Of the about 300 school buildings retrofitted in Nepal, many are located in the areas impacted by the Gorkha earthquake. All the schools that were retrofitted and seismically reconstructed performed very well during the April 25 earthquake and the series of aftershocks. Even those school buildings located near the epicenter region were in a condition of immediate usability. Almost all of the retrofitted school buildings were used during the earthquake response as emergency shelter, warehouse, health posts, or safe offices. This contrasts sharply with poor performance of the "unretrofitted" school buildings, including those located near the retrofitted ones, during the Gorkha earthquake. About 80% of the "unretrofitted" schools were damaged beyond repair. Although a detailed study on each of the schools is yet to be done, it can be simply said that retrofitting did enhance the building resilience significantly and the retrofitting technology successfully passed the test by this earthquake.

5.2.4 Program for Enhancement of Emergency Response
5.2.4.1 General
As an effort toward fulfilment of one of the initiatives of the Kathmandu Valley Earthquake Risk Management Action Plan, NSET started informal training in school on drop, cover, and hold on; and some light search and rescue (SAR) skills to the common people. Later, NSET joined added to efforts in emergency response, first in helping the Asian Disaster Preparedness Centre (ADPC) and later as a leader since 2003 in the implementation of the Program for Enhancement of Emergency Response (PEER), in six countries of Asia including Nepal. PEER, supported by the United States Agency for International Development and the Office of US Foreign Disaster Assistance (USAID/OFDA), is a regional training program that assists beneficiary countries to enhance their emergency response capacities, thereby reducing mortality rates from emergencies and disasters.

5.2.4.2 Program Components
PEER consists in imparting a system of training programs on Medical First Responder (MFR), Collapsed Structure Search and Rescue (CSSR), Community Action for Disaster Response, and Hospital

Preparedness for Emergencies (HOPE). All these training programs are aimed at developing instructors for the respective courses (Tandingan and Dixit, 2012). Therefore, it feeds two parallel streams: (1) training strategy and training curricula for developing the responder's end-users, and (2) training strategy and training curricula for developing national, regional, and international instructors in all of the aforementioned courses. PEER is implemented in collaboration with the Ministry of Home Affairs, government of Nepal (MoHA-GoN, 2013). It has so far trained more than 1100 MFR-CSSR professional responders in Asia (NSET, 2015). Of these, approximately 245 are from the Nepalese Army, Nepal Police, Nepal Armed Police Force, and the Nepal Red Cross Society; all mandated with emergency first response tasks.

In additional to the development and execution of the training curricula and training materials, the program also provides cache of training equipment to the principle emergency response organizations of Nepal: the Nepal Army, the Armed Police Force, and the Nepal Police. As the Nepal Ministry of Home Affairs designated the Nepal Police Academy as the focal partner-training institute for the implementation of MFR and CSSR courses in Nepal, the program distributed sets of CSSR equipment to Nepal Police-Central Police Disaster Response Squadron, to complement the institutionalization of CSSR course.

5.2.4.3 Program Outreach at the Community Level
Efforts were made to best contextualize the training program for the beneficiary country. In addition to imparting state of the art skills on emergency response to national level first responders, the program extended the training to grassroot community groups. In 2014, NSET in collaboration with senior PEER instructors from Nepalese Army, Nepal Police, Armed Police Force, and Nepal Red Cross Society designed and developed the MFR-CSSR curriculum for training end users. This initiative came at the time when Nepal's security forces established their respective Disaster Management units and started to train emergency response teams. The program imparts training on Basic Emergency Response (BEMR), Community Search and Rescue (CSAR), Damage Assessment Trainings (DAT), Vulnerability and Capacity Assessment, and Community Fire Response Training, developing and testing emergency response plans and prepositioning

emergency supplies at different levels (Jimee et al., 2012). So far NSET has provided earthquake preparedness orientation programs for more than 37,000 people and produced more than 1400 CSAR responders, more than 300 BEMR responders, and more than 100 DAT graduates. The target trainees are individuals, communities, government and non-governmental organizations, academic and health institutions, and other international agencies.

5.2.4.4 PEER Program Impact in Development and Enactment of the National Disaster Response Framework

The direct impact of PEER is the enhanced capacity of the country and communities to prepare for emergency response through skills imparted to their members responsible for emergency response. Indirectly, it also serves helping the government develop an emergency response framework. Most of the PEER graduates and instructors belong to emergency response agencies, became catalysts in promoting PEER, designed similar training curricula for their organizations, and delivered similar emergency response training for their response personnel.

In 2013, National Disaster Response Framework (NDRF) was enacted, aiming to develop a clear, concise, and comprehensive disaster response framework for Nepal that can guide a more effective and coordinated national response in case of a large-scale disaster NDRF (2013). The NDRF was endorsed by the government of Nepal to fill the legal, policy, and operational gaps in disaster risk management. This framework has been central in activating the disaster response after the April 25 earthquake in Nepal. The effectiveness of PEER and the role of its graduates played in the response of the 2015 earthquake are discussed later.

5.2.5 Building Code Implementation Program Nepal

5.2.5.1 Building Code Development and Its Implementation Progress

Development of an earthquake loss scenario in the KVERMP and other assessments of seismic risk revealed that the major source of earthquake risk in terms of human casualty in urban areas of Nepal is from the possible collapse of buildings. Effective implementation of a building code for safe construction is one of the most effective ways to reduce the potential risk of casualty from earthquake.

Nepal drafted the NBC in 1994 based on a scientific assessment of seismic hazard and risk. It also considered the then existing practice of building construction including the usage of different construction types and technologies in different climatic regions and the socioeconomic context. The government of Nepal, through the Nepal Building Act of 1998 made the NBC mandatory to be enforced through local governments in municipalities and urbanizing settlements. Most of the early efforts in building code implementation were made though NSET through its program of Nepal earthquake risk management program (NERMP) and subsequently through municipal earthquake risk management program (MERMP).

The first formal enforcement of the code was made in 2003 in Lalitpur submetropolitan city, one of the municipalities in the Kathmandu valley. Since then, several municipalities gradually initiated the enforcement of the building code. Despite efforts made by NSET, other organizations, and the government, only a few municipalities in Nepal are in the process of Building Code Implementation. So far, 50 out of 263 municipalities implemented the NBC. Fig. 5.3 shows municipalities in Nepal implementing the NBC at different levels, ranging from nonenforcement to complete effective enforcement as of 2011. In 2011, NSET started a separate dedicated program, the Building Code Implementation Program in Nepal (BCIPN), which supported 24 municipal governments in public awareness, capacity building and institutional system for code enforcement (Guragain et al., 2017a). Still, the majority of the cities are yet to enforce the code in their building permit process. Lack of institutional capacity and human resources are the main hindrance in the effort. Individual municipalities in the country need a customized approach and process of building code implementation as they vary in terms of their size, use of building construction materials, and building construction process.

5.2.5.2 BCIPN and Progress of Code Implementation After 2011

Considering the need to scale up the good initiatives based on the experience of working from 1998 to 2011 in some selected municipalities, NSET started the BCIPN to support a large number of city governments at once with funding support from the USAID/OFDA. BCIPN envisioned building on and consolidating the experiences, achievements, and lessons learned from earlier similar initiatives. The goal of the BCIPN is to encourage effective compliance to and

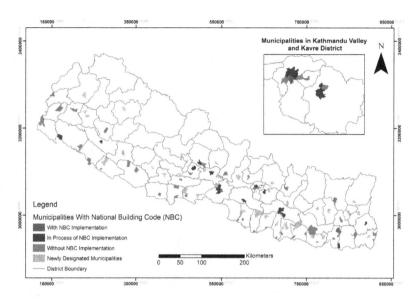

Figure 5.3 Map of Nepal showing municipalities with different stage of building code implementation by the end of 2011.

enforcement of the NBC, taking the approach of forging close partnership between the municipal authorities and other key role players of the society in making their settlements earthquake resilient.

At the beginning of the program, an objective baseline survey of potential municipalities was conducted to understand the existing situation of municipalities in terms of basic information as well as status of building code implementation. Only three municipalities were found implementing the Building Code with effective plans and programs. Nine municipal governments did not even have an engineer in the city office. It was obvious that even though the Building Act of 1998 made code compliance mandatory, a majority of new buildings in most urban and urbanizing areas of Nepal continued to build buildings without of a building permit. Most of the existing buildings lacked any plan or the plans did not conform to the actual built situation.

At the beginning of the BCIPN, a risk perception study was carried out in target municipalities to understand existing level of knowledge, attitude, and practice. The survey showed that people are aware to some level of the earthquake risk and perceived the need for making safer buildings, but it was found that practice of safer construction is almost nonexistent. The study reveals that the major hindrances were lack of an enforcement system at municipalities and insufficient skill on safer construction. Otherwise, people were willing to construct their houses safer. This analysis was helpful to refine specific strategy and approaches BCIPN.

5.2.5.3 BCIPN Strategy for Implementation

Based on a decade-long experience and survey of cities in the areas of urbanization, human resource, intuitional system, and public perception of risk, the BCIPN adopted a three-pronged approach to assist target cities: (1) enhancing earthquake awareness and risk communication, (2) helping technical and institutional capacity building, and (3) institutionalizing the building code implementation process by introducing policy changes. Fig. 5.4 shows the interlinkage of the components of the strategies intended outcomes. Three specific activity components of BCIPN namely, (1) development of an improved building permit system, (2) a code compliance check mechanism, and (3) development of building code implementation guideline and checklist for code compliance align with the program strategy.

5.2.5.4 Institutionalization of Building Code Implementation Through BCIPN

The BCIPN is in the process of being implemented over last 5 years through activities put in strategic steps in each municipality that spreads over the country and has created enabling environment at national level. The program starts from raising awareness in the population that creates demand for housing safety along with capacity building of both human resources and institutional structure. Implementing activities through stakeholders' engagement in a cluster approach and creating a diffusion process to spread over other cities. The success of the program is marked by the fact that governing councils of 50 municipalities have accepted as a policy to enforce the national building code. As the process of institutionalization of building code implementation, these cities have

1. Incorporated building code implementation as an important heading in the annual municipal planning and started allocating a certain budget.
2. Established a separate building code implementation unit within the municipal organization structure, allowing for recruitment of more building engineers.

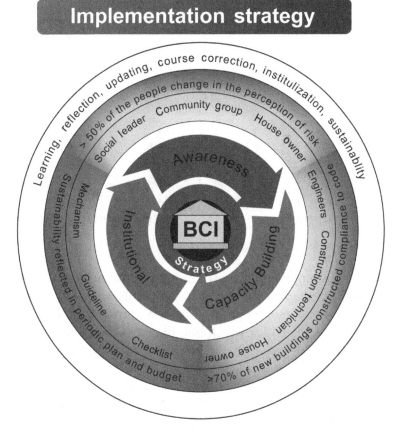

Figure 5.4 Schematic diagram of BCIPN implementation strategy.

3. Started a program for providing orientation on the benefits and process of a building code and building permits for the population, especially for the prospective house owners desiring to construct a new house or repair the old one.
4. Started a registration and licensing system for local builders and masons.

5.2.5.5 Code Compliance in BCIPN Cities
The NSET developed a methodology for evaluating code compliance that is based on study and analysis of the building plans and designs submitted to the municipality as well as field inspection of the actual building constructed on site. The criteria evaluate three major attributes of vulnerability: building configuration in plan and elevation,

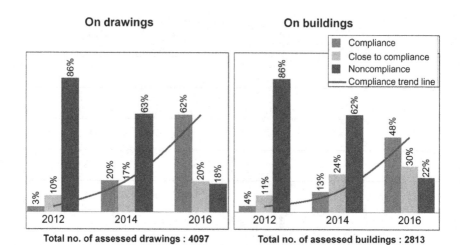

Figure 5.5 *Change in building code compliance in Bharatpur in the period 2012–16.*

strength of the building elements and materials, and the ductility of the structural elements. The results of a compliance survey show significant success of the program. Fig. 5.5 shows a sample case of Bharatpur municipality, in which buildings designed and constructed in three different years (2012, 2014, and 2016) were analyzed. It shows that there is a significant change in compliance expressed in the quality of the drawings submitted along with the building permit applications and, more significant, in the way the building is actually constructed on site. In the sample case presented, the quality of the building plan increased from a compliance rate of 4.2% in 2012 to a compliance rate of 86.9% in 2016. Inspection of buildings on the ground revealed the progression of code compliance from a meager 6.3% in 2012 to a solid 53.8% in 2016 with noncompliance reducing from 81% to 23%. This indicates that city governments in Nepal can actually enhance the seismic performance of new buildings if they are provided with technical assistance for capacity enhancement, improved institutional structure, and proper policy and legal environments.

5.2.5.6 Needs and Gaps in Building Code Implementation

During the period of the BCIPN, the government of Nepal came up with a National Plan of Action for Safer Building Construction and Implementation of Building Code in 2015 involving stakeholders, those who are supporting in the building code implementation.

In confirmation of BCIPN lessons, the action plan identified following problems and challenges:

1. Despite the enactment of Building Act of 1998 and NBC of 2003 along with rigorous efforts of DUDBC, the national building codes and other relevant codes and building bylaws could not be enforced yet throughout the country as envisaged in the act.
2. There is a low level of awareness among the local people and other stakeholders along with a prevailing inadequate capacity of local government institutions, professional engineers, and architects.
3. The coordination, support, and collaboration among various public, private, and local institutions and other stakeholders responsible for implementation of NBC on ground are weak.
4. Effective efforts on identification, mobilization, and allocation of internal and external resources for implementation of NBC have not been made.
5. In addition to reconstruction, retrofitting, and maintenance of earthquake-affected structures, about 1435 new public buildings are constructed annually across the nation. There is a need of systematic code enforcement mechanism to regulation these constructions.
6. Coordination, support, and collaboration among various government institutions, 744 local government levels in seven provinces is a major challenge for successful enforcement of NBC.
7. More than 75% of population is still living in rural settlements, where most of the buildings are constructed with local technology and materials, which may not be compatible with current NBC requirements to withstand the disasters to which they are exposed. There is huge challenge to develop affordable and locally acceptable technological options and material choices for rural housing for different ecological and development regions.

5.3 THE GORKHA EARTHQUAKE: IMPACT AND RESPONSE

5.3.1 General

The Gorkha earthquake was not unexpected. The people of Kathmandu and Nepal were geared toward expecting a large event. However, it was not in the active memory of people and the government.

The devastation was great in rural areas and some urbanizing settlements along the northern axis from the epicenter of the main shock on

April 25, 2015, to that of the largest aftershock on May 12, 2015. In Kathmandu the devastation was well below expectation—partly because of the long period nature of the seismic waves against the low heights of the majority of the buildings, low intensities of shaking, and also the "lucky" timing on Saturday when the schools were closed and people were in their full consciousness. Damage to the critical facilities, including the only international airport, was minimal without significant disturbance to the services.

It was the largest shaking in the country since the 1934 Nepal Bihar earthquake. The impact was huge and people continued to be scared due to the frequent aftershocks. Apart from the social and psychological impacts, the earthquake affected the geological and geotechnical conditions, lifelines, private and public buildings, hospitals schools, and other public services, cultural heritage sites, and historical monuments.

5.3.2 Search and Rescue

In the Gorkha earthquake of 2015, as in many other earthquakes, SAR work was done largely by the community members. More organized SAR operations done by groups of responders belonging to the three national security organizations: the Nepal Army, the Armed Police Force, and the Nepal Police. The Red Cross mobilized huge number of volunteers and trained responders to work with community teams and security forces. NSET also mobilized squads of trained rescuers to assist in the operations of Security Forces in Kathmandu. A total 76 fully equipped and self-sufficient international SAR teams with a total of more than 4300 responders quickly arrived in Kathmandu via the international airport and assisted in SAR operations (EERI, 2016; Nepalese Army, 2015a,b; APF, 2015).

The remoteness of the affected area with difficult terrain, damaged roads, and poor accessibility, as well as adverse weather hindered effective SAR. Lack of SAR equipment was another gap felt by all agencies. The newly drafted national disaster response strategy as well as the series of disaster simulations implemented by the government in collaboration with civil and military organization of friendly countries were very helpful in coordination and understanding the impact. Training of the national responders of Nepal and SAARC countries under the USAID/OFDA-funded and NSET-implemented PEER helped maximize the efforts amid limitations and coordinate efforts with international teams.

5.3.2.1 PEER Reflections in the Gorkha Earthquake

The Gorkha earthquake inflicted huge damage, prompting the government to declare a state of emergency in about a third of the country. The more than 15 years of PEER investment in Asia was put to test during this challenging episode in Nepal's history. The national capacity developed in the past two decades and the national discourse and actions on emergency preparedness were found very handy, albeit "not enough," for responding to the situation.

Nepal PEER partner organizations mobilized response teams for SAR as commanded by the government. PEER-trained professionals provided leadership to the trained or yet-to-be-trained security personnel to undertake SAR operations. Primary PEER partners in Nepal (the Nepalese Army, Nepal Police, Armed Police Force, Nepal Red Cross Society, and NSET) were all in full action. For years, PEER helped embed the basic SAR skills in local and national responders; and because of this foundation, national responders were able to assist the international urban search and rescue (USAR) teams as they had a good understanding of international SAR practices, guidelines, techniques, and the goals of searching and rescuing victims using the safest techniques. It was obvious that the main difference in the work of the national responders' team and the international SAR teams was the difference in access to equipment, mainly the heavy-duty and more sophisticated SAR equipment used by the international teams. Some national response teams worked independently on-site and also assisted the international USAR teams with their understanding of the context and environment.

EERI (2016) report on the performance of the national responders, notwithstanding the problems of logistics, was superb. Had the same level of earthquake happened a decade earlier, the performance of SAR could have been much less effective.

The Gorkha earthquake once again revealed the importance and usefulness of community SAR volunteer responders. It was clear that community volunteers are in fact the first responders and that they needed the training programs listed previously together with the training in first aid. Additionally, the earthquake also revealed the need to train security forces from the private sectors, including those from the tourism and travel business, industry, hotels, river rafting, and so forth, a learning that NSET has been propagating for two decades.

5.3.3 Medical Response

With more than 25 hospitals and 1000 other health facilities destroyed in the worst affected districts, the health cluster under the coordinating leadership of WHO and the Ministry of Health conducted rapid assessments, coordinated deployment of medical teams including 131 foreign medical teams and including several field hospitals and primary care centers, and provision of health services delivery, outbreak surveillance, distribution of essential medicines, and awareness of sanitation and other diseases-prevention measures.

A review 1 year later stated that the health sector's response to the earthquake was rapid, well-coordinated, and tailored to the needs of the affected population. The main lesson learned was that the retrofitting of the Kathmandu hospital and training of its staff in mass casualty management and emergency response plans helped greatly in the background of more than 80% of health facilities in the worst affected districts were destroyed. Strengthening emergency preparedness for disaster response, including preparedness plans and training for all health facilities, should be done at all levels across the country, for which there is a need for stronger policies to ensure DRR measures are implemented and all staff members are trained.

5.3.4 Relief

The Nepal government led the earthquake response largely through the Ministry of Home Affairs (MOHA) of the government of Nepal. Government led the relief and rehabilitation efforts by mobilizing national as well as international resources as stipulated by the Natural Calamity (Relief) Act of 1980. This included among others, (1) financial support to the families with casualties and damaged or lost houses including conceptualization of special loans for earthquake victims, (2) facilitating temporary shelters and relief camps for individuals and the community, (3) conceptualization of relief packages including a National Reconstruction and Rehabilitation Fund (NRRF), (4) development of an early Rehabilitation and Reconstruction Plan backed by the NRRF, (5) coordination of relief efforts by several national and international organizations and benevolent charities, (6) conceptualization of the

National Reconstruction Consultation Committee (NRRC), (7) information management including collection of data and its dissemination, (8) maintenance of law and order, (9) management of the Prime Minister's relief fund, (10) formulation of Post Disaster Need Assessment (PDNA), and (11) coordination of the response by other line ministries including development of indicative list of priority relief items, issuance of an Earthquake Victim Identity Card, concept for deployment of volunteers, and so on. Therefore, the government approached the response in a comprehensive way, leading to the organization of the International Donors' Conference on earthquake reconstruction and rehabilitation in Kathmandu on June 25, 2015.

Responding to the concerns of the national parliament, the government put into effect an Integrated Action Plan for Post-Earthquakes Response and Recovery, 2072. Consisting of forty main items, some of which are listed next:

- Special focus on construction of temporary emergency shelters for special groups of people in the affected areas, such as senior citizens; single women; families with diseased, pregnant, and new mothers; and people with disability.
- Provision of shelters and free education for orphans of earthquake victims.
- Classification of victims for providing support based on the level and nature of impact.
- Provision of free seeds and subsidies for fertilizer and tax exemptions for affected farmers.
- Review, updating, and effective enforcement of NBC of 2006 and the land-use policy.
- Organized demolition of damaged structures.
- Commitment to build a memorial to the Gorkha earthquake victims.

5.3.5 Post Disaster Need Assessment

Very soon after the emergency response was under control, Nepal started a comprehensive assessment of the damages and losses caused by the earthquake to enable planning the earthquake recovery and reconstruction and developing the reconstruction strategy. The PDNA was done with the support of Nepal's long standing development partners, the National Planning Commission of Nepal in collaboration with the government agencies as well as the civil society and nongovernmental organizations, private sector businesses, and international donor agencies operating in Nepal. The PDNA identified recovery needs, developed first cost estimates for these needs, and recommended implementation arrangements. The PDNA took a long-term goal of enhancing national resilience to disasters, promulgated DRR and Build Back Better as the cardinal principles, emphasized raising awareness, and made improvements in national DRR systems in the short, medium, and long terms.

The Ministry of Federal affairs and Local Development (MOFALD) led the Early Recovery process in coordination with MOHA and other agencies of the early recovery cluster. The Earthquake Early Recovery Programme focused on (1) debris management, (2) reconstruction of community infrastructures, and (3) restoration of public service delivery. Emergency livelihoods and economic recovery are taken as a cross-cutting issue that is to be considered the central focus of all activities under these three pillars. The program focused on mobilizing grassroots organizations such as local service providers, civil society organizations, ward citizen forums, and citizen awareness centers under the overall leadership of the district and village level governing bodies. MOFALD developed the following guidelines for early recovery activities:

- appropriateness to local context including cultural context
- long-term sustainability
- directed toward strengthening existing local institutions and mechanisms (VDCs, Municipalities, DDCs, WCFs, CACs, and other local government offices)

For the first time in Nepal, the government could organize post-earthquake building triaging in the Kathmandu valley and some other urban settlements by mobilizing professionals after a brief training.

NSET carried out detailed damage assessment of more than 200,000 earthquake-damaged buildings in Kathmandu and several districts and captured seismic performance of different building types at different intensities of shaking. This yielded lots of information on Nepalese building types, leading to formulation of fragility curves (Guragain, 2017b).

Arranging for temporary housing was another area that the country managed without much controversies. The effort consisted of

(1) mobilizing human resources and financial support; (2) offering technical support, including a definition of standard design and materials standards; (3) providing training for the construction of temporary shelters, including toilets, water supply (water tank, water pipe, water reservoir, maintenance of water supply, assessment of safeness of water), temporary schools and health centers or health posts; (4) provision of human, technical, and material support for the repair and renovation of the partially damaged houses and office buildings; (5) establishing temporary workplaces for the local government institutions, including assessment, repairs, human resources, and office facilities; (6) development and distribution of a temporary ID card for those who lost theirs; (7) continued governance assistance, e.g., by establishing a public hearing desk at DDCs and municipalities and conducting regular public hearings; (8) strengthening of coordination mechanisms among international and national partners including NGOs; (9) ensuring child protection, gender equity and social inclusion, and compliance to international norms for several other cross-cutting issues.

The effort followed proven international practices, such as cash for work, emergency livelihoods and economic recovery, equal opportunities for training for women, on-farm and off-farm activities and establishment of institutional mechanisms and training cash for work, and equal pay for equal work for women in (1) on-farm activities such as vegetable and other short-term farming, support for seeds and fertilizer, renovation of small-scale irrigation and other agro-infrastructures, and training on livestock and vegetable farming; (2) off-farm activities such as fast food production, weaving, bamboo crafts, small groceries, training as masons, plumbing, electrical and electronics good maintenance. It also provided support to national level earthquake recovery planning. Given the large-scale impact of the earthquake in almost half of the country, the recovery efforts would require engagement over a period of at least 2–5 years. Lessons from the previous disasters' response such as the Jure landslide 2014 and the Koshi flood in 2011 show a significant amount of human resources were done rather satisfactorily.

Unfortunately, concept of transitional shelter was not very actively articulated in the society. This led to unrealistic expectations among the population as well as at the political leadership on the possibility of providing shelter before "the next monsoon" or "before the next winter."

5.3.6 Key Lessons Learned From Response to Gorkha Earthquake

The following could be considered as the main lessons learned from emergency response to Gorkha earthquake:

1. Preparedness paid. Preparedness initiatives undertaken by the government agencies were helpful for the government and the people in facing the earthquake and organizing response.

    a. The government could handle more than 60 international SAR teams from different countries. Preparedness efforts toward multinational coordination, civil-military synchronization, simulation planning, and joint disaster drills carried out in the past were instrumental for the state of disaster preparedness of our security agencies—despite serious lack of resources and prior experience in handling disasters of such magnitude, the government could take over effective control and manage the situation adequately. The NDRF and the experience of developing district disaster response plans in almost all the 75 districts and the local disaster risk management plans in several districts and village development committees also helped in this.

    b. Training imparted to the three securities agencies in MFR, CSSR, HOPE, Dead Body Management, and a host of community-based disaster risk management programs by several national and international partners were extremely helpful and prepared the players to home in on the tasks quickly after the earthquake.

2. The cluster system did work. The cluster system, which encouraged preparedness sectorwise helped a lot by quickly activating with response in their respective areas. The shelter cluster could muster the strengths with the members and helped the government in providing advice and emergency action for providing immediate shelter to the population affected. Similarly, the education cluster quickly took stock of the situation and developed concepts of temporary, semi-permanent, and permanent learning centers and helped the government muster resources to construct these so as to restart disrupted education process at the earliest point.

Disaster awareness initiatives implemented in Nepal and the organized efforts of the Nepal government, especially since the UN Yokohama World Conference on Disaster Management; and the country's active implementation of the HFA; the commitment made toward fulfilment of the Sendai Framework for Disaster Risk Reduction (SFDRR); as well as other global initiatives, such as SDGs and climate change adaptation, did help in enhancing people's perceptions of risk and made them at least mentally prepared for impending disasters. During the Gorkha earthquake, people were scared but kept calm and no panic or looting issued, which indicated an enhanced understanding of the hazard and trust in the government among the population had strengthened. Because of a certain level of preparedness, the government started managing the disaster immediately after the event within its capacity. The earthquake did, however, reveal a serious lack of capacity in terms of prepositioning of SAR and other heavy equipment, food and nonfood items, and readiness of a functioning voluntary system.

5.4 RECONSTRUCTION

The earthquakes of April 25 and May 12, 2015, in Nepal affected almost half of the country, including isolated mountainous areas, causing approximately 9000 people to die and more than 22,000 people to be injured. Rapid visual damage assessment, detailed damage assessment, and postdisaster needs assessment were conducted after the earthquake. The government of Nepal formulated the Act Relating to Reconstruction of the Earthquake Affected Structures of 2015, and the National Reconstruction Authority was established. The Post Disaster Recovery Framework (PDRF) was developed and reconstruction started. This section summarizes the lessons from damage assessment, highlights the key strategies of the PDRF, and describes a program for technical assistance to housing reconstruction and its preliminary lessons.

5.4.1 Damage Assessment

The PDNA conducted by the government of Nepal shows widespread destruction of housing and human settlements resulting from nearly 500,000 houses destroyed and more than 250,000 houses partially damaged (PDNA, 2015). The PDNA shows total loss of $7 billion by the earthquake and about $3.5 billion is from housing sector. This indicates that a significant volume of reconstruction is in housing sector and the necessity of a proper understanding of building damage for the formulation of a housing reconstruction strategy.

The main types of buildings in the affected area are stone masonry with mud mortar, some buildings with stone and brick masonry with

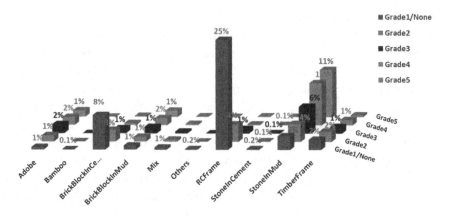

Figure 5.6 Overall damage grade distribution versus building type.

cement-sand mortar and few reinforced concrete buildings with masonry infill. Fig. 5.6 shows the building typology distribution in 31 districts affected by the April 25, 2015, earthquake in Nepal. It shows that about 58% of the buildings are mud-based masonry, i.e., stone in mud, adobe, or brick in mud; 21% are cement-based masonry, either stone with cement-sand mortar or brick with cement-sand mortar; and about 15% are reinforced concrete with masonry infill. Other types of buildings make up only about 6% (CBS, 2011). Among the damaged buildings, about 96% of the buildings were masonry and about 4% reinforced concrete buildings with masonry infill based on preliminary analysis conducted for preliminary damage and needs assessment (PDNA, 2015).

Just after the earthquake, though mainly in the Kathmandu valley, many organizations provided rapid visual assessments of the buildings and suggested people on the possibility to continue using the buildings or whether evacuation was required. However, information on the possibility of repair and retrofitting or demolition was not covered by the rapid visual assessment.

NSET conducted a detailed damage assessment of over 150,000 buildings in 12 municipalities of different locations, with high-intensity to low-intensity impact from the Gorkha earthquake. The survey covered 100% of the buildings in the surveyed locations, and Fig. 5.6 shows the distribution building types versus damage grades.

Damage assessment result shows that the damage was focused in rural area for nonengineered masonry buildings like stone in mud and brick in mud especially in rural areas where there is limited access to engineering professionals.

5.4.1.1 Key Lessons for Reconstruction From Damage Assessment

Following key lessons can be drawn analyzing the damage statistics:

Focus on appropriate technology. Most of the buildings damaged during the Gorkha earthquake was to masonry nonengineered buildings. It is difficult to transport modern construction materials to remote areas. Even in recent years, people are constructing stone-in-mud buildings. So, it is important that the reconstruction techniques should cover stone-in-mud buildings with proper use of locally available materials for disaster resistance.

Technical assistance is key for safer construction. Widespread damage to nonengineered buildings indicates that the construction workers did not know about earthquake-resistant techniques. So, it is recommended to plan a massive training programs and plans for training existing masons. At the same time, new masons may need to be trained, as the need is many times greater for reconstruction than regular construction.

A comprehensive program considering different aspects of technical support is important. The technical support should be comprehensive, covering awareness of home owners about earthquake resistance, the capacity to build a construction workforce, and the support from the local governments to establish a system of compliance checking during reconstruction. The awareness of home owners can help create demand for safer construction, as they control the construction of houses. The trained construction workforce will supply the demand, and the local governments can make the process more sustainable through having the system in place.

5.4.2 Reconstruction Act and Establishment of a National Reconstruction Authority

The Nepal Legislature Parliament enacted the Act Relating to Reconstruction of the Earthquake Affected Structures of 2015 on December 20, 2015 (NRA Act, 2015). The act provided establishment of a National Reconstruction Authority with clear functions, duties, and powers. The act also formed a National Reconstruction Advisory Council with the prime minister as chairman and several members including relevant ministers as members. A steering committee and executive committee with clear duties and responsibilities were also formed.

5.4.3 Post Disaster Recovery Framework

The PDRF was prepared under the leadership of the NRA, in consultation with key stakeholders, aiming to provide a systematic, structured, and priority framework for implementing recovery and reconstruction (PDRF, 2016). It is prepared as a common framework meant to serve all of the government, as well as national and international partners and other recovery stakeholders. The PDRF outlines five strategic recovery objectives that are listed as (PDRF, 2016)

1. Restore and improve disaster-resilient housing, government buildings, and cultural heritage sites, in rural areas and cities.
2. Strengthen the capacity of people and communities to reduce their risk and vulnerability and to enhance social cohesion.
3. Restore and improve access to services, and improve environmental resilience.
4. Develop and restore economic opportunities and livelihoods and reestablish productive sectors.
5. Strengthen the capacity and effectiveness of the state to respond to the people's needs and effectively recover from future disasters.

5.4.4 Baliyo Ghar, a Program of Technical Support for Housing Reconstruction

In the wake of the April 25 earthquake and its subsequent aftershocks, the government of Nepal identified the huge need and demand for trained human resources to complete its goal of "Building Back Safer." The Housing Reconstruction Technical Assistance Program (Baliyo Ghar) program was an initiative to support the government's goal by providing technical support on a wide range of activities implemented by the NSET and supported by the USAID. The objective of the program is to ensure disaster-resilient construction of houses through awareness, training, demonstration, and support for code compliance. The program duration is for 5 years from October 1, 2015, to September 30, 2020.

The "Baliyo Ghar" intends to develop disaster-resilient construction guidelines and training curricula, train the construction workforce, and implement an awareness campaign, leading to standard national resources for training and awareness, enhanced local capacity, and a changed perception of communities toward safety, thereby supporting disaster-resilient earthquake reconstruction and finally contributing to making communities in Nepal disaster resilient.

The following are the main guiding principles for Baliyo Ghar Program:

- country-led policies and processes
- strategic planning, standardized implementation, and involvement of all stakeholders
- inclusion and access
- integrating DRR
- flexibility and context specific approaches

With these guiding principles, the program is assisting mainly in three areas of work:

1. *Support on policy formulation and development of curricula and guidelines at the central level.* The program supports development of fund disbursement guidelines and developing information booklets, posters, and pamphlets related to the grant distribution. Baliyo Ghar contributed in preparing a standard operating procedure for building inspection; a manual for inspection, technical posters incorporating 10 Tips for Earthquake Resistance, development of a 7-day masons training curricula for urban and rural masons, and so forth.
2. *Capacity development.* The program trained more than 1100 engineers as potential trainers, more than 4000 masons trained on earthquake-resistant construction, and trained about 200 social mobilizers who continue providing orientation to people.
3. *Awareness and house-to-house support.* The program reached to about 50,000 people through direct orientation on the need and possibility for resilient construction.

5.4.5 Preliminary Lessons From Reconstruction
After one and half years of implementation, some issues are noted as important lessons for consideration in the coming years:

- There is 10%−20% of reconstruction of housing one and half years after the start of reconstruction. The compliance rate is much higher in the area where the comprehensive technical assistance is provided than in the areas where there is no technical assistance.

- Technical consultation and awareness activities at the household level are key for safer construction, as they help sensitize house owners on need for and importance of safer construction. Before the consultation, many people thought that earthquake-resistant construction using local construction materials was not possible. Those who heard and had knowledge of earthquake-resistant technologies believed that such construction requires huge investments. Through door-to- door campaigns and orientation programs, it was possible to change the mind set of the community.
- Demonstration models are found as proof of the concept of "Seeing is believing."
- Scarcity of construction materials and escalation of price took place in some locations. For example, people are demanding and using more timber from the community forest for construction, and conservation offices are controlling it. Alternative technologies need to be explored and promoted.

In remote villages, where there are no roads to transport construction materials, the local community has been facing technical challenges due to unavailability of construction materials for earthquake-resistant component.

5.5 FUTURE DISASTER RISK REDUCTION FOR NEPAL

5.5.1 General

The efforts of past two decades toward disaster risk management in Nepal have demonstrated positive impacts during the 2015 Gorkha earthquake. Several initiatives in the past helped Nepali communities to suffer less damage during the earthquake, and several initiatives were extremely useful for effectively responding to it with immediate SAR, relief, and recovery. The foundation created helped to move toward better and safer reconstruction. The following are key aspects to consider and to continue in the future in a more organized, systematic, and institutionalized manner.

1. *Continue, scale-up, and institutionalize successful initiatives.* Initiatives such as the SESP, Building Code Implementation Program, and PEER have clearly shown their usefulness and

impacts during the earthquake. However, their coverage so far is limited. For example, the school retrofitting program implemented as part of SESP has reached to about 300 schools; however, there are more than 80,000 school buildings throughout the country. More than half the school buildings are highly vulnerable to earthquakes. The school retrofitting program needs to reach out to all schools in the country. This requires more scaling-up efforts with more institutionalized and organized approaches. Likewise, the building code implementation program has so far reached less than 50 municipalities of the country; however, there are now 744 total local governments in Nepal. Such building code efforts should reach to all the local communities to ensure safer building construction everywhere. Similarly, the emergency response capacity enhancement has been limited to national level response organizations at the central level only. Emergency response capacity is needed at all villages and municipalities; therefore, the capacity building programs should reach to lowermost levels. The institutionalization process of such efforts already is taking place somewhat. This needs a more accelerated pace.

2. *Continue enhancing the awareness and capacities of all stakeholders.* Increased awareness and capacities of stakeholders are essential for creating demand and subsequently fulfilling the need and demand of earthquake safety. Development of awareness and capacity are required at all levels: from policy and decision levels at the center to the implementation levels at the communities. Significant progress has been achieved during the past decades in raising awareness and building capacities. Future disaster risk management efforts should continue emphasize raising awareness and building capacity.

3. *Expand from urban areas to rural areas.* Efforts of the past decade covered mostly the urban areas of Nepal. The successful initiatives should now be replicated to the rural areas as well.

4. *Bring successes and lessons of reconstruction to entire country.* Although the pace of reconstruction is slow as compared to the aspirations of the affected population, the current policies, institutional system, and implementation framework has laid a strong foundation for sustainable and safer reconstruction. The process for ensuring safer building construction through technical support and a compliance inspection mechanism has been established in the earthquake-affected areas. However, such a system is not yet established in many other areas, urban as well as rural. The need now is

to replicate the positive lessons and successes of the reconstruction to the whole country. Only 14 out of the 75 districts were badly hit and 17 other districts were affected by the 2015 Gorkha earthquake. The positive influence of the reconstruction process is largely concentrated in the 14 badly hit districts only. Now, the need is to bring the lessons from 14 to 31 and then to all 75 districts of the country.

5. *Mainstream DRR issues at political levels.* Despite the tremendous progress in the past decades toward DRR and preparation, the agenda of DRR has not yet become a main priority for a disaster-prone country like Nepal. It has not attracted enough attention and has not received the required priority at the political levels. Hence, the institutional setup as envisioned by the National Strategy for Disaster Risk Management (NSDRM), formulated and approved by the government in 2009, has not yet been formed. NSDRM envisioned a high-powered authority, National Authority for Disaster Risk Management (NADRM) at the central level, for looking after the overall disaster risk management in the country (NSDRM, 2009). Likewise, similar powerful authorities at the subnational and local levels have also been proposed. The priority for future should be to develop such focused institutions at all levels, so that the efforts of disaster risk management can be made more efficient and effective.

6. *Exploit the opportunities provided by global and regional initiatives.* The global and regional efforts, such as the HFA: 2005–15 and the SFDRR are some key concepts endorsed during the global conferences. These concepts have helped significantly to frame Nepal's DRR policies and plans (NSDRM, 2009). Likewise, initiatives and platforms, such as the GPDRR, the AMCDRR, the Global Earthquake Mode, are other opportunities from which Nepal can get tremendous benefits in terms of institutionalizing better policies.

5.6 CONCLUSIONS

Earthquake risk management initiatives in Nepal in the past two decades have yielded good results: Earthquake awareness increased across the country as never before; use of awareness initiatives, such as knowledge transmission using the vast network of community FM radio stations and engagement of common persons in such state-led initiatives as the annual ESD on January 15, have contributed much to the achievement. Fatalism has gradually given way to scientific belief across the layers of the society. However, this change has not yet reached the desired critical proportion for sustaining the change process independently. Efforts of DRR in Nepal should now transgress its emphasis from awareness raising to implementation of vulnerability reduction initiatives, including stopping the creation of vulnerability. Now is the time to enhance risk perception through demonstrating the benefits of risk reduction by actually implementing risk reduction methodologies—Nepal has enough success stories; the need is to scale up the successes in DRR.

The vulnerability of residential buildings constitutes the major source of seismic risk measured in terms of human casualties during earthquakes. A significant proportion of commercial, and public sector buildings are also feared as vulnerable to earthquakes. More than two-thirds of earthquake risk comes from poorly constructed buildings without in-built earthquake resistance. This is largely due to the large stock of existing buildings that were constructed in the past 3–4 decades, even before the advent of the national building code. Unfortunately, many of the new construction, in urban as well as in rural areas, continue to be built without proper compliance to the largely earthquake-centric national building code. Reducing earthquake risk in Nepal largely means inculcating the practice of making buildings to withstand the earthquake forces. Nepal is blessed with a very rich tradition of constructing earthquake-resistant structures as witnessed in the cultural heritage sites strewn all around the Kathmandu valley and in many historic rural settlements all over the country. Effective implementation of the national building code in urban as well as rural areas is one of the most effective ways to reduce the potential risk of casualty and loss of assets from earthquakes.

Three core issues identified for proper implementation of the building code at the municipal level are raising awareness, building capacity, and institutional setup. Process on building code implementation suggested a sequence as (1) building awareness and capacity, (2) engaging stakeholders, (3) providing pilot implementation to some municipalities, (4) scaling up and implementing processes at large scale, (5) sharing lessons and experiences and enhancing the system.

The building code implementation helped institutionalize the DRR activities at the local government level. The major lessons learned are

(1) raising awareness and building capacity are the keys to successful institutionalization of DRR at the municipal level; (2) institutionalization of DRR in the local governance is a long-term process and needs persistent efforts; (3) identification of champions and working under their leadership is the best policy to accelerate the institutionalization of DRR; and (4) a critical threshold of achievement needs to be crossed for an en masse acceptance of the new concept, such as mainstreaming DRR as business usual in all aspects of life. This requires continuation of the efforts and increased investment by the nation as well as by the donor community in the field of disaster risk management. The window of opportunity has been opened up by the 2015 Gorkha earthquake sequence and the process of proper reconstruction will help internalize DRR into the livelihood of the people, not only in the earthquake-affected areas but also in areas yet to be shaken by the next earthquake.

REFERENCES

ADB/GON, 2010. Concept Paper for Vulnerability Reduction of Schools in Kathmandu Valley.

Adhikari, L.B., Gautam, U.P., Koirala, B.P., Bhattarai, M., Kandel, T., Gupta, R.M., et al., 2015. The aftershock sequence of the 2015 April 25 Gorkha–Nepal earthquake. Geophys. J. Int. 203 (3), 2119–2124.

Adhikari, S.R., Dixit, A.M., Guragain, R., Murakami, H., 2016. Study on Shaking Intensity Distribution of the 2015 Gorkha Earthquake in Nepal. In: International Workshop on Gorkha Earthquake "Lesson Learned and Future Road Map for Safer Community and Sustainable Development" 24–25 April 2016, Organized by Government of Nepal.

Armed Police Force (APF), 2015. APF Operation Search and Rescue in Earthquake 2015, Role of PEER Certified graduates from NSET, Nepal.

BCDP, 1994. Building Code Development Project: Seismic Hazard Mapping and Risk Assessment for Nepal; UNDP/UNCHS (Habitat) Subproject: NEP/88/054/21.03. Ministry of Housing and Physical Planning, Kathmandu.

BCPR/UNDP, 2004. United Nations Development Programme. Bureau for Crisis Prevention, & Recovery. Reducing Disaster Risk: A Challenge for Development-A Global Report. United Nations.

Bordet, P., Colchen, M., Le Fort, P., 1972. Some features of the geology of the Annapurna Range Nepal Himalaya. Himalayan Geol. 2, 537–563.

CBS, 2011. National Population and Housing Census 2011. Nepal.

Dixit, A.M., Dwelley-Samant, L.R., Nakarmi, M., Pradhanang, S.B., Tucker, B.E., 2000. The Kathmandu Valley Earthquake Risk Management Project: an evaluation. In: Proceedings World Conference on Earthquake Engineering (12WCEE), Auckland, 2000.

DMG, 2002. Seismic Hazard Mapa of Nepal (1:1,500,000). Available from Department of Mines and Geology as reported in < www.seismonepal.gov.np >.

Dunn, J.A., Auden, J.B., Gosh, A.M.N., Roy, S.C., 1939. The Bihar-Nepal earthquake of 1934. Mem. Geol. Surv. India 73, 391 pp.

EERI, 2016. Earthquake Reconnaissance Team Report: M7.8 Gorkha, Nepal Earthquake on April 25, 2015 and Its Aftershocks. <https://www.eeri.org/2016/05/nepal-earthquake-reconnaissance-team-report-is-now-available/> (assessed June 2017).

Fuchs, G., Sinha, A.K., 1988. The tectonics of the Garhwal-Kumaun lesser Himalaya. Jahrb. Geol. Bundesant. (Austria) 121 (2), 2019−2241.

Gansser, A., 1981. The geodynamic history of the Himalaya. In: Gupta, H.K., Delany, F.M. (Eds.), Zagros, Hindu Kush, Himalaya; Geodynamic Evolution, Geodynamics Series, vol. 3. Amer. Geophys. Un., Washington, DC, pp. 111−121.

GESI, 2001. A Report on Global Earthquake Safety Initiative Pilot Project Final Report, Implemented by GeoHazards International (GHI) AND UNCRD, 86 p. Available in: <http://www.preventionweb.net/files/5573_gesireport.pdf> (accessed 01.12.12.).

Guragain, R., 2017b. Building Damage Patterns Of Non-Engineered Masonry And Reinforced Concrete Buildings During April 25, 2015 Gorkha Earthquake In Nepal Proceedings World Conference on Earthquake Engineering (16WCEE), Santiago Chili.

Guragain, R., Pradhan, S.P., Maharjan, D.K., Shrestha, S.N., 2017a. Building code implementation in Nepal: an experience on institutionalizing disaster risk reduction in local governance system. In: Shaw, R., Izumi, T., Shiwaku, K. (Eds.), Science & Technology in Disaster Risk Reduction in Asia: Potentials and Challenges 1st Edition, Elsevier, ISBN:9780128127117.

Hagen, T., 1969. Preliminary Reconnaissance. Report on the Geological Survey of Nepal 86. Denkschriften der Schweizerischen Naturforschenden Gesellschaft, vol. 1. 185 pp.

Heim, A., Gansser, A., 1939. Central Himalaya Geological Observations of Swiss Expedition, 1936. Hindustan Publishing Corporation, Delhi, 246 pp.

Jimee, G.K., Upadhyay, B.U., Shrestha, S.N., 2012. Earthquake awareness programs as a key for earthquake preparedness and risk reduction: lessons from Nepal. In: The 14th World Conference on Earthquake Engineering.

Le Fort, P., 1975. Himalayas, the collided range: present knowledge of the continental arc. Am. J. Sci. 275A, 1−44.

Le Fort, P., 1996. Evolution of the Himalaya. In: Yin, A., Harrison, T.M. (Eds.), Tectonic Evolution of Asia. Cambridge University Press, New York, NY, pp. 95−109.

Liu, G., Einsele, G., 1994. Sedimentary history of the Tethyan Basin in the Tibetan Himalaya. Geol. Rundsch. 83, 32−61.

Ministry of Home Affairs-Government of Nepal (MoHA-GoN); National Strategic Action Plan on Search and Rescue 2013; August 2013.

Nakata, T., 1982. A photogrammetric study on active faults in the Nepal Himalaya. J. Nepal Geol. Soc. 2, 67−80.

Nakata, T., 1989. Active faults of the Himalaya of Nepal and Nepal. In: Malinconico, L.L., Jr., Lillie, R.J. (Eds.), Tectonics of the Western Himalaya. Geol. Soc. Amer., Spec., Pap. 232. pp. 243−264.

NDRF, 2013. National Disaster Response Framework. Government of Nepal, Ministry of Home Affairs, Kathmandu. Available from: < www.ifrc.org/docs > .

NSET, 2015. NSET PEER Experiences, In-house report on PEER Country Planning Meeting, Nepal.

NRA Act, 2015. Act Relating to Reconstruction of the Earthquake Affected Structures, 2015 (2072). Ministry of Law and Justice, Government of Nepal.

Nepalese Army, 2015a. Life Turning to Normal in Devastating Earthquake Affected Districts by the 'Sankat Mochan' operation of Nepalese Army <http://www.nepalarmy.mil.np/sankatmochan/view-news> (posted on 16.06.15.).

Nepalese Army, June 2015b. Deployment of PEER Graduates, Nepal.

NSDRM, 2009. National Strategy for Disaster Risk Management. Government of Nepal Ministry of Home Affairs, Nepal. Available from: <http://un.org.np/sites/default/files/report/2010-08-06-nsdrm-in-eng-2009.pdf>.

NSET, 1999a. Kathmandu Valley's Earthquake Scenario. NSET, Kathmandu.

NSET, 1999b. The Kathmandu Valley Earthquake Risk Management Action Plan. NSET, Kathmandu.

NSET-KVERMP, 2010. ADB/Amod Mani Dixit/ Surya Prasad Acharya-Report on National Workshop on School Safety.

NSET, 2016. Disasters in Nepal: Inventory of Events and Analysis of Impacts 1971–2016, June 2016, Lalitpur (in house report).

Pandey, M.R., Tandukar, R.P., Avouac, J.P., Lavé, J., Massot, J.P., 1995. Interseismic strain accumulation on the Himalayan Crustal Ramp (Nepal). Geophys. Res. Lett. 22 (7), 751–754.

Pandey, M.R., Chitrakar, G.R., Kafle, B., Sapkota, S.N., Rajaure, S.N., Gautam, U.P., 2002. Seismic Hazard Map of Nepal. Department of Mines and Geology, Kathmandu.

PDNA, 2015. Nepal Earthquake 2015, Post Disaster Needs Assessment, in two volumes. Government of Nepal, National Planning Commission, Kathmandu.

PDRF, May 2016. Nepal Earthquake 2015, Post Disaster Recovery Framework Government of Nepal National Reconstruction Authority Kathmandu.

Rajaure, S., Asimaki, D., Thompson, E.M., Hough, S., Martin, S., Ampuero, J.P., et al., 2016. Characterizing the Kathmandu Valley sediment response through strong motion recordings of the 2015 Gorkha earthquake sequence. Tectonophysics. Available from: https://doi.org/10.1016/j.tecto.2016.09.030.

Rana, B.J.B., 1935. Nepal Ko Maha Bhukampa (Great Earthquake of Nepal). Jorganesh Press, Kathmandu.

SAARC, 2009. SAARC regional cooperation on earthquake risk management in South Asia: road map. Outcome of the SAARC Workshop on Earthquake Risk Management in South Asia, Islamabad, Pakistan, 8-9 October 2009. <http://saarc-sdmc.nic.in/pdf/roadmap/road_map5.pdf>.

Sapkota, S.N., Bollinger, L., Klinger, Y., Tapponnier, P., Gaudemer, Y., Tiwari, D., 2013. Primary surface ruptures of the great Himalayan earthquakes in 1934 and 1255. Nat. Geosci. 6, 71–76.

Schelling, D., Arita, K., 1991. Thrust tectonics, crustal shortening and the structure of the far eastern Nepal Himalaya. Tectonics 10, 851–862.

Seeber, L., Armbruster, J., 1981. Great detachment earthquakes along the Himalaya arc and long-term forecasting in earthquake prediction: an international review. Maurice Ewing Series 4, 259–279.

Seeber, L., Armbruster, J.G., Quittmeyer, R.C., 1981. Seismicity and continental subduction in the Himalayan arc. In: Zagros, Hindu-Kush, Himalaya, Geodynamic Evolution, Geodyn. Ser. 3. pp. 215–242.

Srivastava, P., Mitra, G., 1994. Thrust geometries and deep structure of the outer and lesser Himalaya, Kumaon and Garhwal (India): implications for evolution of the Himalayan fold-and-thrust belt. Tectonics 13, 89–109.

Tandingan, M.R., Dixit, A.M., 2012. Experiences on Implementing Program for Enhancement of Emergency Response (PEER) in six countries of South East Asia. In: Proceedings World Conference on Earthquake Engineering (14WCEE), Lisbon, 2014.

UNDP/UNCHS, 1994. Seismic hazard mapping and risk assessment for Nepal. Prepared by Beca Worley International consultant in association with Golder Associate, TAEC Consultant, SILT Consultant and Urban Regional Research.

CHAPTER 6

Improving the Nepalese Building Code Based on Lessons Learned From the 2015 M7.8 Gorkha Earthquake

Uddhav Karmacharya[1], Vítor Silva[2], Svetlana Brzev[3] and Luís Martins[2]

[1]Understanding and Managing Extremes Graduate School, Pavia, Italy [2]GEM Foundation, Pavia, Italy [3]IIT, Gandhinagar, Gujarat, India

6.1 INTRODUCTION

Nepal is situated in the seismically active Himalayan region, with five major tectonic zones, including Terai, Siwaliks, Lesser Himalaya, Higher Himalaya, and Tibetan-Tethys Himalaya. Each zone is characterized by its own lithology, tectonics, structures, and geological history (Chaulagain et al., 2015). These tectonic zones are separated by major thrust faults, capable of generating large magnitude earthquakes ($M > 8$) and causing widespread destruction. The thrust faults along the Main Frontal Thrust and Main Boundary Thrust are the most active faults in the region (Lava and Avouac, 2000). The continental collision between the Indian and Eurasian plates has been occurring at a relative rate of 40−50 mm per year (USGS, 2015), and it has been a source of several major earthquakes in Nepal. Chitrakar and Pandey (1986) and GEER (2015) reported several major earthquakes in 1255, 1260, 1408, 1681, 1767, 1810, 1823, 1833, 1834, 1934, 1980, 1988, 2011, and 2015.

The great 1934 Bihar-Nepal earthquake (M_w 8) with the maximum Modified Mercalli Intensity (MMI) of X devastated the eastern part of Nepal, and caused almost 8500 fatalities. More than 50% of the total building stock was either completely destroyed (19%) or heavily damaged (38%) (Rana, 1935). A few decades later, on August 21, 1988, the eastern part of Nepal was struck by an earthquake with a moment magnitude (M_w) of 6.6, which led to 721 fatalities and 6553 injuries.

Impacts and Insights of the Gorkha Earthquake. DOI: http://dx.doi.org/10.1016/B978-0-12-812808-4.00006-7

In total, 66,541 buildings either collapsed or suffered extensive damage (Thapa, 1988).

More recently, Nepal experienced a devastating earthquake of magnitude 7.8 (M_w) with an epicenter at Barpak, Gorkha district, on April 25, 2015, at 11:56 a.m. local time. The earthquake was followed by a strong aftershock of magnitude (M_w) 7.2 on May 12. These events caused 8790 fatalities and more than 22,000 people were injured. It was estimated that 8 million people, almost one third of the total population, were affected by the earthquake (EERI, 2016). Out of 75 districts, 31 were affected and 14 districts were declared "crisis-hit" for the purpose of prioritizing rescue and relief operations. More than 700,000 buildings, including UNESCO World Heritage sites, either collapsed or experienced severe damage. In terms of housing, the majority of the affected dwellings were rural unreinforced stone masonry buildings located in the central and western region of the country (NPC, 2015). The Department of Urban Development and Building Construction (DUDBC) reported a peak ground acceleration (PGA) in Kathmandu of 0.15 g, which is significantly lower than the code-prescribed design PGA of 0.32 g, corresponding to the 300-year return period. In spite of a relatively low ground shaking intensity, reinforced concrete (RC) buildings with masonry infills sustained damage or total collapse at some localities within the Kathmandu valley and other districts of Nepal.

This study explores the key factors influencing seismic performance of RC frame buildings with masonry infills in Nepal. Building damage surveys performed after the 2015 earthquake showed that the amount of masonry infill walls, quantified in the form of wall index (WI), is one of the key factors influencing the performance of low-rise RC buildings (up to three stories high) in the earthquake. Damage observations and results of the building surveys are presented in the study. Several fragility models examining the effect of the variable WI in the Nepalese seismic hazard context were developed using nonlinear dynamic analyses. The results of these analyses were used to develop recommendations for updating the seismic design provisions for low-rise RC buildings in the Nepal Building Code (NBC).

6.2 THE REINFORCED CONCRETE BUILDING PORTFOLIO IN NEPAL

6.2.1 Building Portfolio and Reinforced Concrete Construction Practice in Nepal

Nepal is located in the middle of the Himalayan belt bordered by China on the north and by India on the south, east, and west. It consists of three main geographical regions: the Himalayas is the least inhabited region covering around 15% of the total land area, whereas the Mid-Hills cover about 68% of the total land and hosts 43% of the total population. The third region (Terai) covers only 17% of the total land area but hosts 50% of the population (Chaulagain et al., 2015). The distribution of the buildings within these districts is similar to the population distribution. The spatial distribution of the buildings in Nepal is illustrated in Fig. 6.1. The National Housing Census Survey of 2011 documented that RC buildings represent about 18% and 10% of the total building stock in urban areas in the Terai region and Kathmandu valley, respectively. Around 80%–90% of the total building stock in Nepal is composed of mud mortar- or cement mortar-bonded brick and stone masonry buildings, wooden and adobe buildings (CBS, 2012). At the national scale, 44% of the building stock is composed of mud cement-bonded stone or brick masonry (BM/SM); 18% are cement mortar-bonded brick or stone masonry structures (BC/SC); 25% are composed

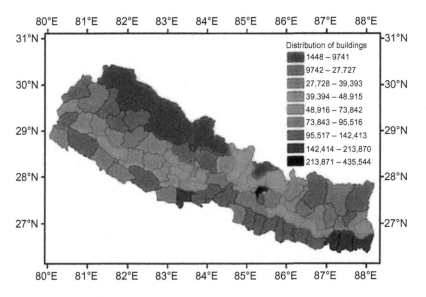

Figure 6.1 Districtwise distribution of buildings in Nepal.

wood buildings (W); 10% are RC construction (RCC); and only 3% are adobe buildings (A). The vast majority of the buildings are nonengineered, that is, constructed using informal construction techniques, and as a result they exhibited poor seismic performance when exposed to earthquake ground motions. In fact, these informal construction techniques were the main reason for the unacceptably high human and economic losses in the 2015 earthquake.

The practice of RC construction in Nepal started in the late 1970s. Over the last four decades, the rate of construction of RC buildings has been rapidly increasing, and these buildings have replaced older buildings of adobe, stone, and brick masonry construction (Chaulagain et al., 2013). In the 1990s, almost half (49%) of all new buildings constructed in Nepal were reported to be of RCC, while in the 1970s RC buildings constituted only 11% of the new construction. Due to rapid construction growth and a lack of consistent code enforcement in the country, RC buildings are often constructed by petty contractors without following seismic provisions. In other words, these buildings can be considered nonengineered buildings. It was reported that the structures designed in compliance with the code provisions (well-designed structures) account for only 8% of the total RC building stock (Dixit, 2004; Shrestha and Dixit, 2008).

Densely populated urban and semiurban areas of Nepal are usually dominated by two- to five-story RC buildings with brick masonry infills, but the majority of the buildings is three-stories high. Typical low-rise RC buildings in Nepal are shown in Fig. 6.2. Many buildings have mixed-functions, with a ground floor used for commercial activities and upper floors used for residential purposes. These buildings, known as *storefront buildings*, have one or two open sides in their plan (see Fig. 6.2A,B). RC floor and roof structures are typically composed of slabs with a thickness of 100 mm. A postearthquake survey of buildings damaged in the 2015 earthquake (Brzev et al., 2017) has shown that the typical column size was 230 mm square. Beams in these buildings are typically 230 mm wide, while the depth ranges from 305 to 425 mm. RC columns and beams typically had four or more longitudinal deformed steel bars (of variable sizes), while the transverse reinforcement (ties) were usually in the form of 7 mm diameter bars at 200 mm spacing (in some cases 5 mm wires were also observed). In majority of the buildings, tie anchorage was provided by means of 90°

Figure 6.2 Typical low-rise RC buildings in Nepal: (A) open storefront buildings in Kathmandu; (B) a two-story building with top story columns for future expansion; (C) a building under construction, and (D) a completed residential building. From S. Brzev and U. Karmacharya.

hooks. Masonry infill walls were built using burnt clay bricks in cement mortar. It was observed that the exterior walls are thicker (230 mm) than the interior walls (115 mm).

6.2.2 Seismic Design Provisions for Reinforced Concrete Buildings According to the Nepal Building Code

One of the most efficient mechanisms to mitigate the impact from earthquakes is the enforcement of rigorous seismic design regulations (Spence, 2004). The Nepal National Building Code (NBC) was developed by the DUDBC of the Ministry of Physical Planning and Works (MPPW) in 1993, with the assistance of the United Nations Development Programme (UNDP) and United Nations Centre for Human Settlement (UN-HABITAT). NBC was enforced when the Building Construction System Improvement Committee (established by the Building Act of 1998) authorized the MPPW (UNCRD, 2008). The code was targeted to the most common building types in Nepal,

such as RC buildings. It did not consider the sophisticated design phi-
losophies and analytical techniques that can be found in the building
codes of developed countries, but NBC 105:1994 provides basic seismic
design provisions. In 2010, the DUDBC published additional recommendations for the construction of Earthquake Safer Buildings in
Nepal with the assistance of UNDP (UNDP, 2010). These guidelines
are an improvement of the NBC, in particular concerning the recommendations for the minimum size of columns for buildings up to three-
stories high, minimum reinforcement for various structural elements,
limitation of the beam length, and the like.

Three levels of sophistication are related to the design and construction of RC buildings. *Professionally engineered structures* represent those
RC buildings that are designed and constructed with adequate seismic
provisions and ductile detailing in compliance with the Indian Standard
IS 13920:1993. *Preengineered design* for RC buildings of restricted size
(up to three-stories high) is addressed by Mandatory Rules of Thumb
(MRT). Sizes of the key structural components, reinforcement details,
and standard design drawings are included (NBC 201:1994). *Guidelines
for nonengineered masonry construction*, such as stone masonry housing
prevalent in rural areas, are also available (NBC 203:1994).

The MRT present ready-to-use diagrams and detailing of key
structural components for the design of RC buildings with a plinth area
up to 93 m^2, which are up to three-stories high (or 11 m overall height)
and of regular plan shape and elevation (no setbacks). Neither the
length nor the width of the building can be more than 25 m, the bay
length cannot be greater than 4.5 m, and the length of the building cannot be greater than 3 times the width. The MRT recommends the size of
a beam to be 230 × 325 mm, whereas two recommendations regarding
the column size can be found. The column size on the ground floor
abutting infill walls should be 230 × 300 mm and all other columns
should have a section of 230 × 230 mm. The shear reinforcement in
beams is in the form of 6 mm diameter stirrups at 150 mm spacing,
while for the columns 8 mm diameter ties at 100 mm spacing are
adopted. No special provisions are provided for beam-column joints
and the effects of masonry infill are not considered in the design. The
section size of beams and columns and reinforcement details are presented in Tables 6.1 and 6.2. It can be seen from Table 6.2 that MRT
implicitly establishes the minimum required column area per floor.

Table 6.1 Reinforcement Details of Beams

Beams	Story No.	Support NBC/NBC+	Midspan NBC/NBC+
	1	2ø16 + 2ø12 / 2ø16 + 2ø12	2ø16 / 2ø16
	2	2ø16 + 2ø12 / 2ø16 + 2ø10	2ø12 / 2ø16
	3	3ø12 / 3ø12	2ø12 / 2ø12
	1	3ø16 / 3ø16	2ø16 / 2ø16
	2	2ø16 + 2ø12 / 2ø16 + 2ø12	2ø16 / 2ø16
	3	3ø12 / 3ø12	2ø12 / 2ø12
	1	3ø16 / 3ø16	2ø12 / 2ø12
	2	2ø16 + 1ø12 / 2ø16 + 1ø12	2ø12 / 2ø12
	3	3ø12 / 3ø12	2ø12 / 2ø12

The MRT have been compulsory in all municipalities and some village development committees in Nepal since 2003. The Lalitpur Sub-Metropolitan City was the first municipality to enforce the NBC, including the MRT.

6.2.3 Seismic Failure Mechanisms for Reinforced Concrete Frames With Masonry Infills

The seismic response of RC frame systems with masonry infill walls has been the subject of numerous experimental and analytical studies since the 1950s. Many buildings of this type have been exposed to damaging earthquakes around the world, which provided an additional opportunity to observe damage patterns and the key parameters influencing their seismic response. When subjected to earthquake ground

Table 6.2 Section Size and Reinforcement Details of Columns

Columns	Story No.	Cross Section of Column
		NBC
	1	4ø16 230 × 300
	2	4ø12 230 × 230
	3	4ø12 230 × 230
	1	4ø16 230 × 300
	2	4ø12 230 × 230
	3	4ø12 230 × 230
	1	8ø12 230 × 230
	2	4ø12 230 × 230
	3	4ø12 230 × 230

shaking, these buildings may experience either a flexural- or shear-dominant behavior. A desirable flexural failure mechanism is characterized by the development of flexural hinges in RC columns and beams. The construction of ductile RC frame systems requires advanced construction skills and adequate supervision. Past earthquakes in several countries have confirmed that achieving ductile flexural behavior in an earthquake may be a challenge, particularly in countries without proper code enforcement. Collapse of RC buildings has led to significant fatalities during past earthquakes in several countries, including the 2001 Bhuj, India, earthquake and the 1999 Turkey earthquakes (Murty et al., 2006).

Alternatively, RC frames with masonry infills can experience a shear failure, which is characterized by shear failure of masonry infill walls and adjacent RC columns. This failure mechanism may take place when there is a strong masonry infill in combination with a nonductile RC frame (Martín Tempestti and Stavridis, 2017). An infill panel can be characterized as "strong" based on the relative stiffness of infill and adjacent RC columns. It can be assumed that, in low-rise buildings, the stiffness is shear dominant and that its flexural component can be ignored. Furthermore, shear failure of a RC column, which is a prerequisite for a shear-failure mechanism, will take place provided that its shear capacity is less than the shear force corresponding to the column's flexural capacity (which corresponds to the flexural failure mechanism). These criteria were established based on the experimental and analytical studies on RC frames with infills subjected to reversed cyclic loading (Mehrabi et al., 1994; Martín Tempestti and Stavridis, 2017). A recent experimental research study on nonductile RC frames with infills showed that a test specimen that experienced shear failure (Basha and Kaushik, 2016) had a damage pattern similar to that observed in RC buildings affected by the 2015 Gorkha earthquake (see Fig. 6.3).

6.2.4 Wall Index
In RC frames with masonry infills and a shear-dominant failure mechanism, the lateral load-resisting capacity of the system depends on

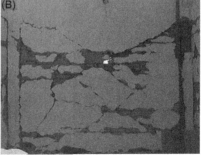

Figure 6.3 Shear failure of RC frames with infills: (A) experimental study (Basha and Kaushik, 2016), and (B) observed damage after the 2015 earthquake. From S. Brzev.

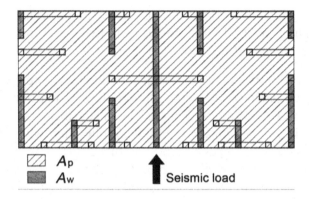

the shear capacity of the masonry walls. In fact, both the lateral load-resisting capacity of a RC frame with infills and its stiffness depend on the cross-sectional area of an infill wall. For that reason, a WI (also known as *wall density*) can be used to assess the lateral load capacity of the building for a specific direction of earthquake shaking. Several studies have demonstrated a strong correlation between the WI and the extent of earthquake damage for masonry buildings in Chile (Moroni et al., 2000, 2004). Several studies in Turkey (Hassan and Sozen, 1997; Gulkan and Sozen, 1999) explored a correlation between the wall and column indices and extent of damage in RC buildings affected by the earthquakes in that country.

The WI for a specific direction in a building can be determined as a ratio of the sum of the cross-sectional wall area in a given earthquake shaking direction, A_w, and the total floor plan area, A_{ptotal}, as expressed in Eq. (6.1) (e.g., Brzev et al., 2017). WI is determined at the base level of the building (ground-floor level), where seismic demand is largest, thus A_{ptotal} denotes the sum of floor plan areas above the base of the building (see Fig. 6.4):

$$\text{WI} = \frac{A_w}{A_{\text{ptotal}}} \qquad (6.1)$$

Note that A_w is computed as the product of the wall thickness and length (where the length may also take into account the dimensions of adjacent RC columns). It is important to note that an infill wall should not be considered in the WI calculation in the following cases:

1. Infill walls with openings, in which the area of the unconfined opening is greater than 10% of the wall surface area of a wall panel enclosed by RC columns.
2. Infill walls characterized by the height-to-length aspect ratio greater than 1.5 (which are not expected to have a significant shear capacity).

It is very important to emphasize that WI can be used to assess seismic safety of walled buildings with a regular plan shape and wall layout. This method cannot take into account torsional effects (without modification) or irregularity over building height. It is assumed that walls are continuous over the building height; that is, discontinuous walls should not be taken into account in the WI calculations. This approach also assumes that all walls at the story level simultaneously reach their shear strength.

It is possible to estimate the required WI for a building with a given seismic hazard level, type of soil, masonry shear strength, expected seismic performance (ductility), average story weight, and the number of stories (Meli et al., 2011). The required WI value can be derived by stating that the required lateral load-resisting capacity of a building at its base needs to exceed the seismic demand (seismic base shear force).

In addition to the WI, column index (CI) has been used in research studies on seismic vulnerability of RC frames with masonry infills (e.g., Hassan and Sozen, 1997; Gulkan and Sozen, 1999). The CI value is related to the building plan and column layout irrespective of the earthquake direction, and it can be determined as the sum of cross-sectional areas for all RC columns and the total floor plan area A_{ptotal} (same as used for the WI calculations). Note that the WI and CI definitions may be somewhat different in different research studies, however the basic concepts are the same.

A relationship between the WI and the damage grade of RC frame buildings with masonry infills that were affected by the 2015 Gorkha earthquake has been explored in this study, as discussed in the following sections.

6.3 EARTHQUAKE DAMAGE OBSERVATIONS

6.3.1 General Observations

In spite of the relatively low level of ground shaking (PGA of 0.15 g in Kathmandu), several RC buildings were affected by the earthquake. The damage patterns ranged from minor damage (cracks in the masonry walls and RC columns) to complete collapse of several buildings in Kathmandu and smaller communities located closer to the epicenter (e.g., Dolakha and Sindupalchok districts). It should be noted that severely damaged RC buildings in Kathmandu were found at a few localized areas (pockets), and the damage was due to the higher intensity ground shaking at those locations (unfortunately the ground motion records were not available). It is important to note that buildings with similar construction features did not suffer any damage at some other locations in Kathmandu. Some RC buildings were reported to have experienced permanent ground displacement, implying a foundation failure (EERI, 2016). Buildings on sloped ground experienced more severe damage than buildings in flat soil areas. Reconnaissance studies of RC buildings were performed by several groups (Brzev et al., 2017; Gautam et al., 2016; Guragain et al., 2017; EERI, 2016). It is believed that the main causes of earthquake-induced structural damage in low-rise RC buildings were (1) inadequate detailing of RC structural components and poor construction quality, (2) increased seismic demand due to structural irregularities causing extensive damage at the ground-floor level, and (3) shear or flexural failure of RC frames with infills.

The soft-story mechanism was the most common failure mechanism found in the collapse of RC buildings with mixed commercial and residential functions (Gautam et al., 2016). The ground-floor level in this type of building, used for commercial purposes, was either left almost open or provided with only a few masonry infill walls. This lack of stiffness at the ground-floor level led to a large displacement demand, thereby causing the soft-story mechanism. Fig. 6.5A shows a five-story building that had a restaurant on the ground floor and a setback at the top floor level (Brzev et al., 2017). When the ground floor collapsed, the building moved by more than 2 m away from the adjacent building, which remained undamaged in the earthquake (see Fig. 6.5B). It was observed that the longitudinal reinforcement (made from Torkari

(A)   (B)

Figure 6.5 A soft-story collapse of an RC building in Sitapaila, Kathmandu: (A) an exterior view of the building showing the collapsed ground floor; (B) a side view showing the direction of collapse. From S. Brzev.

steel) in one of the columns fractured due to significant tensile stresses at the base of the building.

A few common reinforcement detailing flaws were observed in the majority of the damaged buildings, including (1) excessively wide tie spacing in RC columns (200 mm or more), as shown in Fig. 6.6A; (2) use of column ties with 90° hooks (instead of 135°); and (3) lap splices in the column longitudinal reinforcement at floor locations (and inadequate splice lengths) (see Fig. 6.6B). It was also observed that ties were not provided in beam-to-column joint areas.

A flexural failure mechanism (weak column–strong beam) was observed in a few damaged RC buildings, and it likely caused the collapse of several other buildings. This mechanism can be illustrated on an example of a four-story building with an open storefront in the Sitapaila area of Kathmandu, as shown in Fig. 6.7A. Flexural hinges formed at the ground-floor level, both at the base (Fig. 6.7B) and the top of a column (Fig. 6.7C). It is also apparent that the column detailing was deficient (i.e., insufficient transverse reinforcement), which caused buckling of the longitudinal reinforcement (Brzev et al., 2017).

(A)   (B)

Figure 6.6 Inadequate RCC and detailing: (A) excessively wide tie spacing and 90° anchorage in RC columns and (B) lap splices provided at the floor level and an absence of ties at the beam-column connection. From B. Pandey.

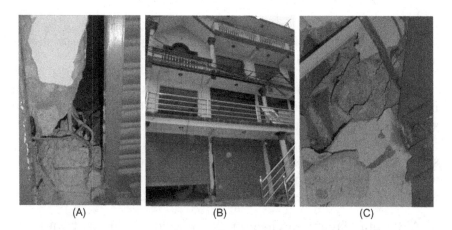

(A)   (B)   (C)

Figure 6.7 A building with an open storefront with the weak column–strong beam failure mechanism: (A) a façade view showing hinges at the top and bottom of a ground-floor column, (B) a flexural hinge at the base of the column, and (C) a flexural hinge at the top of the column. From S. Brzev.

An extensive postearthquake survey of approximately 79,000 buildings in the most affected districts of Nepal (Guragain et al., 2017) included 11,838 RC frame buildings with masonry infills (accounting for about 15% of all surveyed buildings). A large majority (92%) of all the surveyed buildings were up to three-stories high. The survey showed that the most common damage pattern in RC buildings was

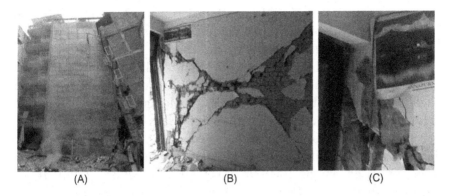

Figure 6.8 Shear failure of an RC frame building with masonry infills in the Gongobu area of Kathmandu: (A) an exterior view of the building in longitudinal direction, (B) an interior transverse wall showing in-plane diagonal shear failure of masonry walls, and (C) diagonal cracks extended into the top of the left RC column shown in (B). From S. Brzev.

damage to infill walls. In fact, 89% of all the surveyed RC buildings had severely damaged infills (damage extent greater than 2/3). It is expected that these buildings showed shear-dominant seismic response characteristic of low-rise nonductile RC buildings with strong infills, as discussed earlier in this chapter. The shear-failure mechanism in RC frames with infills can be illustrated in an example of a six-story building in the Gongobu area of Kathmandu that experienced heavy damage at the ground-floor level (see Fig. 6.8A). Shear failure was observed in several walls at the ground-floor level, and the cracking extended into the adjacent RC columns (Fig. 6.8B,C). It was observed that the quality of RCC was poor, and that the reinforcement detailing was inadequate (Brzev et al., 2017).

6.3.2 Postearthquake Surveys of Damaged Buildings

After the 2015 earthquake at least two detailed surveys of low-rise RC frame buildings were performed. A survey of 98 RC buildings (referred to as BS 1) was carried out by a Canadian and Nepalese team in July 2015 (Brzev et al., 2017). Another survey of 146 RC buildings (BS 2) was conducted in June 2015 by a team from ACI Committee 133, Disaster Reconnaissance (Shah et al., 2017).

Group BS 1 (Brzev et al., 2017) surveyed 98 RC buildings at three localities (Sitapaila, Balaju, and Batar) where significant damage of RC buildings was reported. Sitapaila and Balaju are urban areas located in the Kathmandu valley (about 80 km away from the epicenter of the April 25, 2015, earthquake), while Batar is a suburban area located in the Nuwakot district (about 55 km from the earthquake epicenter). The survey focused on the damage assessment of two- to five-story RC buildings with masonry infill walls, of which 60% were three-stories high and an additional 38% buildings were two-stories high. All surveyed buildings were RC buildings with brick masonry infill walls. Most buildings were characterized by a regular (usually rectangular) plan shape. The authors proposed a revised damage classi-fication for RC buildings with a shear-dominant failure mechanism, which is based on the EMS-98 scale (Grünthal et al., 1998). It was assumed that RC columns experience predominantly shear damage while RC beams do not experience any significant damage due to the nature of this failure mechanism. Damage grades ranged from DG1 (negligible to slight damage) to DG5 (destruction). Out of all the sur-veyed buildings, most buildings (around 80%) experienced damage classified as DG1 and DG2, about 20% buildings experienced damage classified as DG3, and only 3% buildings experienced damage classi-fied as DG4.

An electronic data survey form was developed in the framework of the Global Earthquake Model (GEM) for use with the OpenQuake platform (Silva et al., 2014). Each building is characterized by its loca-tion (latitude and longitude), and 13 attributes describing the details of the lateral load-resisting system, materials, height, shape of the build-ing plan, type of floor/roof, and the like according to the GEM Building Taxonomy V 2.0 (Brzev et al., 2013). The research team also took physical measurements of building Plan dimensions and wall and column dimensions. Multiple earthquake damage photographs were taken for each building.

The south façade of a typical building surveyed in Sitapaila, Kathmandu is shown in Fig. 6.9A (note an open storefront), and its ground-floor plan is shown in Fig. 6.9B. The plan dimensions (12 m × 8 m, length × width) are typical for the surveyed buildings. It was found that the average area of the ground-floor plan for the sur-veyed buildings was 70.3 m^2, with a standard deviation (STD) of 22.04 m^2. All exterior walls were 230 mm (one-brick) thick brick masonry while all interior walls were 115 mm (half brick) thick. Typical RC columns were 230 mm^2. This building was irregular in ele-vation: The top floor had a terrace that covers approximately 25% of

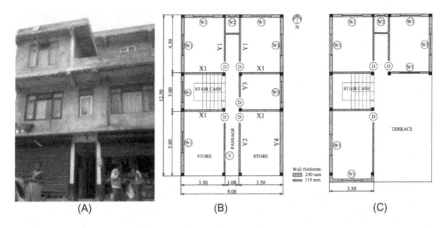

Figure 6.9 An example of a surveyed building in Kathmandu: (A) a photo showing the south façade (entrance), (B) ground-floor plan, and (C) top floor plan (Brzev et al., 2017).

the plan area, and the south wall at the same level was offset with regards to the lower floors (see Fig. 6.9C).

Group BS 2 (Shah et al., 2017) performed a survey of 146 low-rise nonengineered RC buildings, up to seven-stories high. Several field measurements were taken during the survey, including column and wall dimensions and wall locations. Photo logs were used to document the damage. Structural damage was classified as "collapse," "severe," "moderate," or "light." Out of all the surveyed buildings, 46% either collapsed or sustained severe damage, 45% sustained light damage, and the remaining 9% sustained moderate damage. Only buildings up to three-stories high from BS 2 were considered in the current study (36 buildings in total).

An average WI for all buildings (98 in total) in BS 1 survey was 1.38%, with a STD of 1.01% and a coefficient of variation (COV) of 0.703. The WI in the surveyed buildings ranged from 0.19% to 5.65%. Fig. 6.10 shows WI versus the damage grade for all the surveyed buildings and a trend line (solid line). A significant fraction of the surveyed buildings (46 out of 98) were located in Sitapaila, a neighborhood in the capital, Kathmandu. The trend line shows a lower average *WI* than the overall building sample (dashed trend line). The buildings in Sitapaila experienced significant damage compared to other localities in Kathmandu. This can be explained by a lower WI value and the poor quality of RCC observed in Sitapaila.

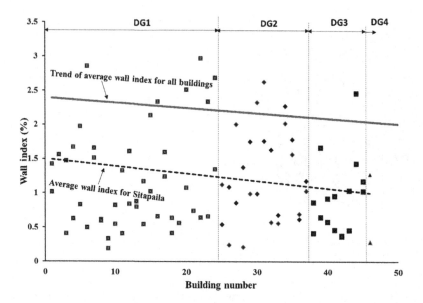

Figure 6.10 *WI versus number of buildings (survey BS 1). Note* solid trend line *showing all surveyed buildings (sample, 98 buildings) while* dashed trend line *shows the results for Sitapaila, Kathmandu (sample, 46 buildings) (Brzev et al., 2017).*

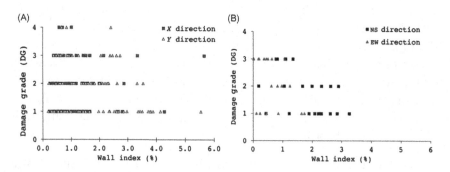

Figure 6.11 *WI versus damage grade.*

The results of both building surveys showed similar average WI values. The average WI values were found to be 1.38% and 1.32% for surveys BS 1 and BS 2, respectively. Fig. 6.11A illustrates how surveyed buildings under BS 1 are clustered into Damage Grades (DG) 1 to 4, depending on their WI; note that WI values are shown in both horizontal directions for each building. The chart shows that the cluster with DG1 is spread over a wide range of WI values, whereas buildings with other damage grades are clustered within smaller WI ranges.

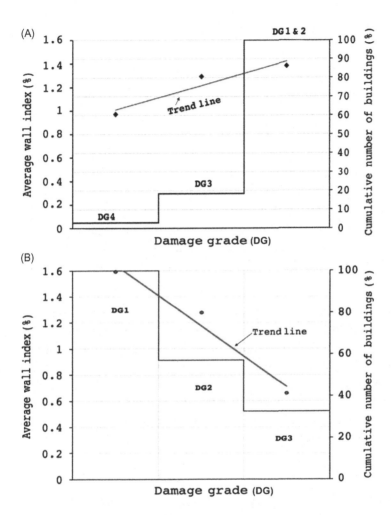

Figure 6.12 WI versus damage grade (cumulative).

An almost identical trend was observed for the results of BS 2, as shown in Fig. 6.11B.

Fig. 6.12 shows a relationship between the cumulative number of damaged buildings characterized by different damage grades and the corresponding average WI value for each cluster. It can be seen from Fig. 6.12A that, for survey BS 1, only 3% buildings experienced DG4 and their average WI was 0.97%. About 20% of all surveyed buildings experienced DG3 and DG4 and the corresponding average WI was 1.3%. Most buildings (about 80% of all the surveyed buildings)

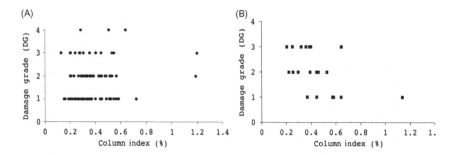

Figure 6.13 CI versus damage grade.

experienced DG1 and DG2, and the corresponding average *WI* was around 1.5%. Similar results were obtained processing data for survey BS 2. Both trend lines indicate a strong relationship between the WI and the damage grade.

However, a relation between the CI value and the damage grade was found to be weak (see Fig. 6.13). It appears that the CI values were clustered over the same range for all damage grades (with a few exceptions). It is interesting that average CI values obtained from both surveys are similar: 0.37% and 0.49% for BS 1 and BS 2, respectively. Similarly, the relationship between WI and CI is also weak; therefore, the authors concluded that CI values in the Nepalese low-rise RC buildings are very low and therefore do not influence their seismic response.

6.4 ASSESSMENT OF THE SEISMIC VULNERABILITY

As empirically demonstrated in the previous section, the WI can significantly influence the seismic performance of a building. This hypothesis has been analytically investigated through the development of fragility functions, which will be presented in this section. A typical three-story building (9 m overall height) chosen for the analyses is shown in Fig. 6.14. The building is an RC frame structure with brick masonry infill walls that are continuous along the building height. The building plan has an area of 96 m^2 with an overall length of 12 m, and 8 m width. The plan consists of three bays, 4 m long each in the longitudinal direction, and three bays in the transverse direction (two spans of 3.27 m and one span of 1.23 m). The building has RC floor and roof slabs, with a thickness of 127 mm.

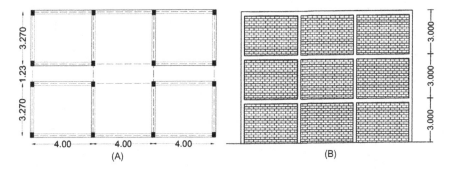

Figure 6.14 Typical layout of case study building: (A) plan view, (B) elevation (all dimensions in meters).

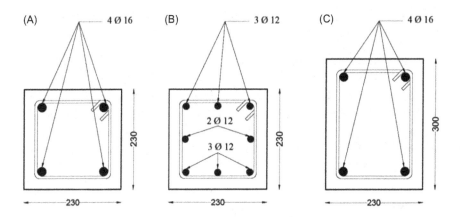

Figure 6.15 Column cross-sections: (A) columns at the first and second floor levels, (B) interior columns at ground-floor level, and (C) exterior column at ground-floor level abutting the infill wall.

The geometrical properties and the reinforcement detailing of the beams and columns commonly adopted in the buildings constructed following the MRT (NBC 201:1994) are presented in Tables 6.1 and 6.2. Typical cross-sections of RC columns and beams are shown in Figs. 6.15 and 6.16, respectively. Transverse reinforcement (ties) in the RC columns was in the form of 7 mm diameter bars at 150 mm spacing. The spacing was uniform along the column height.

The M15 concrete grade (characterized by 15 MPa cube compressive strength at 28 days) was used for RC elements. Fe250 steel grade (minimum characteristic yield strength of 250 MPa) was used for beam stirrups and column ties, while Fe415 grade (minimum characteristic

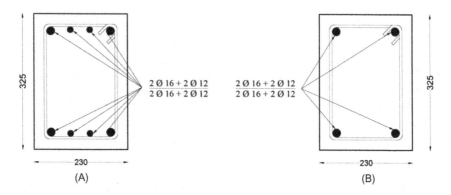

Figure 6.16 Typical beam cross-sections: (A) at end supports and (B) at midspan.

yield strength of 415 MPa) was used for longitudinal reinforcement in beams and columns.

In this study, masonry infills are considered structural elements that significantly influence the seismic response of the structure, as quantified through the WI. Masonry properties were assumed based on the results of past experimental studies from Nepal (Pradhan, 2009). The following properties were used in the study: diagonal compressive strength 0.585 MPa, shear strength 0.28 MPa, and a modulus of elasticity of 2300 MPa.

6.4.1 Numerical Modeling of Reinforced Concrete Frames With Masonry Infills

The case study building was modeled using the finite element software SeismoStruct V7.0.3 (SeismoSoft, 2015). The structural elements were modeled with inelastic force-based frame fiber elements (infrmFB) with five integration points. A uniaxial nonlinear constant confined model for concrete is based on the constitutive law proposed by Mander et al. (1988). The cyclic rules for the confined and unconfined concrete used in the model were proposed by Elnashai and Elghazouli (1993) and Martinez-Rueda (1997). Confinement provided by the transverse reinforcement was taken into account through the rules proposed by Mander et al. (1988). The reinforcement was modeled using the uniaxial model proposed by Menegotto and Pinto (1973).

The masonry infill wall was represented by a 2D Strut model first developed by Crisafulli (1997), and later implemented by Smyrou et al. (2011). The model consists of six strut elements, which allows modeling

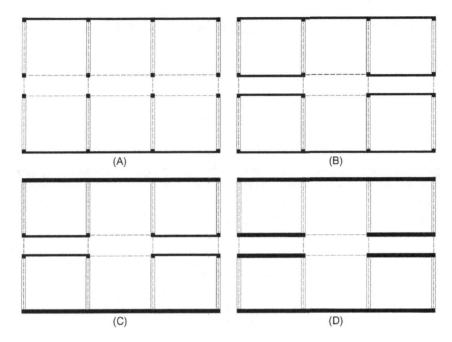

Figure 6.17 Layout of building plans for NBC models showing different masonry infills in longitudinal direction: (A) NBC_IF-1, (B) NBC_IF-2, (C) NBC_IF-3, and (D) NBC_IF-4.

the inelastic behavior of masonry. The struts along the diagonal direction are modeled to resist axial stresses (i.e., compression/tension), while the remaining two struts are provided to resist shear stresses by activation in the direction of compression. Each type of strut is characterized by a dedicated hysteretic model.

6.4.2 Numerical Models: Key Parameters
The main objective of this study is to correlate the WI with the seismic vulnerability of RC frame buildings with masonry infills. For that reason, four structural models (referred here *NBC models*) with different WI values in a longitudinal plan direction (ranging from 0.95% to 3.25%) were considered in this study. It should be noted that the damage surveys discussed in Section 6.3 indicate that an average WI value found in the damaged buildings was on the order of 1.3%–1.4%; which is similar to the average WI value for models NBC_IF-1 and NBC_IF-2 (1.28%). The floor plans for these models are shown in Fig. 6.17 and the WI values are listed in Table 6.3. The WI values were calculated from Eq. (6.1) assuming 115 mm (half brick) and

Table 6.3 Numerical Structural Models and the Corresponding WI Values	
Models	**WI (%)**
NBC_IF-1	0.95
NBC_IF-2	1.60
NBC_IF-3	2.50
NBC_IF-4	3.25

Table 6.4 Definition of the Damage States (Crowley et al., 2004)	
Structural Damage Band	**Description**
DL	Linear elastic response, flexural- or shear-type hairline cracks (<1.0 mm) in some members, no yielding in any critical sections.
SD	Member flexural strength achieved, limited ductility developed, crack width reaches 1.0 mm, and initiation of concrete spalling. Strain limits may be taken as $\varepsilon_{c\,(SD)} = 0.004 - 0.005$, $\varepsilon_{s\,(SD)} = 0.010 - 0.015$
NC	Significant repair required, wide flexural or shear cracks, and buckling of longitudinal reinforcement may occur. Strain limits may be taken as $\varepsilon_{c\,(NC)} = 0.005 - 0.010$, $\varepsilon_{s\,(NC)} = 0.015 - 0.030$ (inadequately confined members).
Complete collapse (CC)	Repair of the structure not feasible either physically or economically, demolition may be required due to shear failure of vertical elements or excessive displacement.
where $\varepsilon_{c\,(SD)}$ = *Strain in concrete at significant damage limit state,* $\varepsilon_{s\,(SD)}$ = *Strain in steel at significant damage limit state,* $\varepsilon_{c\,(NC)}$ = *Strain in concrete at near collapse limit state,* $\varepsilon_{s\,(NC)}$ = *Strain in steel at near collapse limit state.*	

230 mm (full brick) thick infill walls. Note that the effect of openings was not considered in this study.

6.4.3 Definition of Damage States

For the purpose of fragility assessment, three limit states have been defined, as proposed by Priestley (1997) and Calvi (1999): Damage limitation (DL), significant damage (SD), and near collapse (NC). The material strains for each limit state (as defined by Crowley et al., 2004) are presented in Table 6.4. Nonlinear static (pushover) analyses were performed to define the thresholds for different damage states. These thresholds were employed in pushover and nonlinear dynamic analyses to classify the structure into a given damage state.

6.4.4 Nonlinear Static (Pushover) Analysis

Nonlinear static (pushover) analyses were performed to evaluate the seismic capacity for each of the aforementioned models. In this process, the numerical model of the structure is subjected to increasing lateral displacements (drift) in one direction until it reaches the collapse state. During the analysis, the key engineering demand parameters (e.g., roof drift, interstory drift) and the corresponding base shear capacity were recorded. Given the damage observed after the 2015 Gorkha earthquake and the preliminary structural analysis from this study, it is clear that RC frame buildings with masonry infills in Nepal, which were observed to lack ductile design and detailing, could experience one of the following failure mechanisms: a soft-story collapse or a shear failure of the RC frame with infills. Both mechanisms were explained in Section 6.3.1 and illustrated in examples from the 2015 Gorkha earthquake. A shear-failure mechanism was also explained in Section 6.2.3, since that mechanism is the prerequisite for using WI as a measure of seismic vulnerability in RC frame buildings with masonry infills. A soft-story collapse mechanism is more likely to occur in buildings with an open ground floor. The models considered in this study have continuous walls over the building height and are thus likely candidates for a soft-story collapse mechanism.

The capacity curves for the aforementioned four NBC models plus a reference bare frame model (NBC_BF) obtained as a result of the pushover analyses are presented in Fig. 6.18. The analysis results show that the base shear capacity for RC frames with infills is almost 1.5 to 2 times the capacity of an otherwise similar bare frame model. Since the masonry infills are unreinforced walls they are stiff and brittle, which gives the stiff prepeak response. It can be noticed from Fig. 6.18 that the maximum base shear strength is largest for the model with highest WI value of 3.25% (NBC_IF-4). It can be also noticed that, unlike the other three models, the NBC model with the lowest WI value of 0.95% (NBC_IF-1) does not show a drop in the base shear strength due to infill failure. Due to the brittle nature of the unreinforced masonry, the descending branch of the capacity curve after the infill failure takes place is steep. After the masonry infills at the ground-floor level have failed completely, the capacity drops to that of the bare frame. The masonry infills may be damaged in the ground-floor level while they remain in the elastic range in the upper floors. The damage survey of the 2015 earthquake also supports this finding.

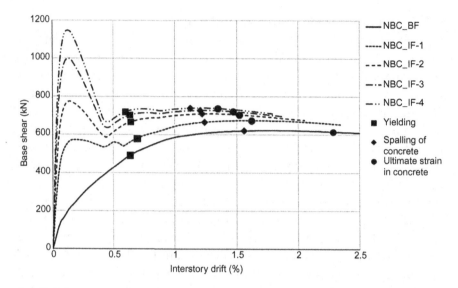

Figure 6.18 Capacity curves obtained from pushover analysis for the four NBC models.

The presence of the unreinforced masonry infills in the building increases the base shear capacity but reduces the displacement capacity. This depends on the WI level: The higher is the WI value, the lower is displacement capacity. The maximum interstory drift for each damage state was estimated based on the results of the pushover analyses and the damage criteria defined in Table 6.4. The results of the pushover analysis have shown that the masonry infills experience either severe damage or collapse at interstory drift of 0.6%.

6.4.5 Nonlinear Dynamic Analysis and Fragility Derivation

The fragility function is a statistical tool representing the probability of exceeding a damage state (performance level) as a function of engineering demand parameter that represents the ground motion. The nonlinear time history dynamic analyses were performed using SeismoStruct v7.0.3 (2015), and the fragility functions were derived through the multiple stripe analysis (MSA), by following the Jalayer and Cornell (2002) approach. A set of 200 ground motion records from the Pacific Earthquake Engineering Research (PEER) Center database were selected, considering the seismicity and tectonic environment around the Kathmandu valley. For all four NBC models, 10 bins (each containing 20 records) were selected from intensity measure of 0.1–2.0 g at an interval of 0.2 g. The maximum interstory drift for all models was captured during each iteration. The plot of maximum interstory drift versus the increasing spectral acceleration (S_a) is presented in Fig. 6.19 for models NBC_IF-1 and NBC_IF-4.

The fragility functions were derived from the MSA results by treating every exceedance of a limit state in a statistical manner. For each intensity measure (IM), the probability of the structure reaching a defined damage state was taken as the fraction of analysis causing drift that exceeds its limit state. Continuous functions were derived by regression analysis of the set of points obtained in this way. A lognormal distribution was assumed, with logarithmic mean and logarithmic STD determined using the maximum likelihood method. The fragility functions defined in terms of spectral acceleration for a period of vibration of 0.5 s for the four NBC models are shown in Fig. 6.20, while the statistic parameters are listed in Table 6.5.

The influence of the WI level upon the probability of damage or collapse can be easily observed on the fragility curves for the four NBC models. The increase in the WI value increases the stiffness and strength of the structure. For example, at spectral acceleration of 1 g (which roughly corresponds to the design PGA for Kathmandu of 0.32 g for stiff rock sites), the NBC_IF-1 model (with the lowest WI value) has a probability of near collapse state of approximately 40%, while the NBC_IF-4 (the highest WI value) has a probability of collapse less than 10%. For a spectral acceleration of 0.5 g (which roughly corresponds to the PGA that was recorded in Kathmandu during the 2015 Gorkha earthquake, 0.15 g), the NBC_IF-1 model has a probability of DL state of 40% and SD of 15%. The results of survey BS 1 discussed in Section 6.3.2 showed that about 20% of all surveyed buildings (which had an average WI of 1.3%) experienced DG3 per EMS-98 scale (this could be correlated to the SD grade). At the same time, only 3% of all surveyed buildings experienced DG4, which is close to collapse. These buildings had an average WI value of 0.97%, which is almost identical to model NBC_IF-1. It can be seen from the fragility curves for that model that the probability of collapse is very low (close to zero) at S_a of 0.5 g, which is in line with the field observations.

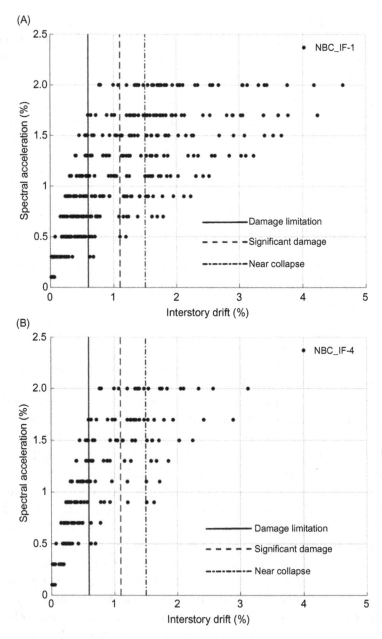

Figure 6.19 The results of MSA analyses showing spectral acceleration versus interstory drift: (A) NBC_IF-1 model and (B) NBC_IF-4 model.

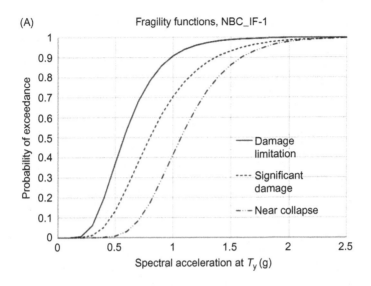

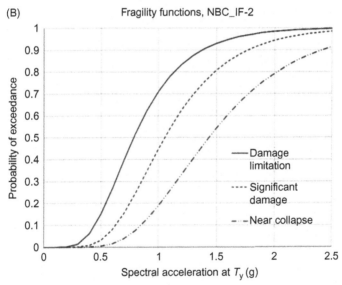

Figure 6.20 Fragility functions in terms of S_a at 0.5 s for NBC models: (A) NBC_IF-1; (B) NBC_IF-2; (C) NBC_IF-3, and (D) NBC_IF-4.

6.5 PROPOSED WALL INDEX FOR NEPAL

From the damage observations of the 2015 Gorkha earthquake (see Section 6.3), it is clear that the WI plays a critical role in the seismic performance of the low-rise RC buildings in Nepal and is potentially more influential than the CI, which is implicitly prescribed by the

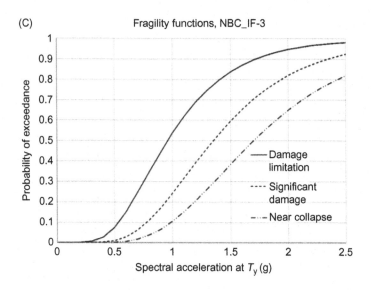

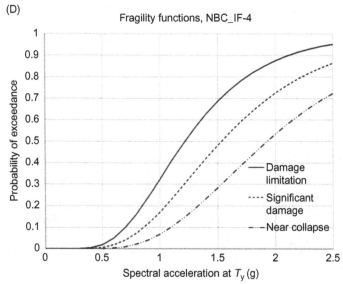

Figure 6.20 (Continued).

MRT. Similarly, fragility functions depicted in Fig. 6.20 also demonstrate a strong correlation between WI and the probability of significant damage or collapse. Such findings can support the improvement of the current code by imposing, in addition to a minimum column area, a minimum WI value. However, given that the seismic hazard

NBC Model	Fragility Models					
	DL		SD		NC	
	λ	ξ	λ	ξ	λ	ξ
NBC_IF-1	−0.56	0.42	−0.22	0.42	0.05	0.31
NBC_IF-2	−0.24	0.44	0.06	0.40	0.35	0.42
NBC_IF-3	−0.04	0.46	0.30	0.43	0.52	0.42
NBC_IF-4	0.20	0.43	0.44	0.45	0.70	0.47

Table 6.5 Lognormal Mean (λ) and STD (ξ) of the Fragility Functions

varies considerably across the territory of Nepal (e.g., Thapa and Guoxin, 2013), it is important to estimate the required WI values for building sites characterized by different hazard levels. An excessively low WI value might lead to insufficient seismic safety, while an overly high WI could signify unnecessarily high construction costs. A possible answer to this dilemma can be obtained by applying a risk-targeted philosophy, in which the seismic capacity or the design ground shaking is defined based on a target safety level, which can be defined by the local government and other decision makers. For example, in the United States the ASCE (2010) regulation established an acceptable collapse probability of 1% in 50 years, which roughly corresponds to an average annual collapse probability (AACP) of 2.0×10^{-4}. In Europe, Silva et al. (2014) proposed an AACP equal to 5.0×10^{-6}, based on an evaluation of several proposals from countries such as France, Denmark, the United Kingdom, and New Zealand. For the case of Nepal, due to the high seismic hazard, it might be unfeasible, and in many cases even technically impossible, to maintain a level of safety similar to that proposed for Europe. For this reason, for the purposes of this study, an AACP equal to 1.0×10^{-4} (i.e., half of what is currently accepted in the United States) has been adopted.

The assessment of the annual probability of sustaining a given level of damage requires a characterization of the seismic hazard across the region of interest, usually represented by a seismic hazard curve (i.e., probability of exceeding a set of ground shaking levels). For this study, the probabilistic seismic hazard analysis (PSHA) model proposed by Thapa and Guoxin (2013) for Nepal was employed. Using the OpenQuake-engine (Pagani et al., 2014; Silva et al., 2014), a set of hazard curves were calculated for the top 13 cities in Nepal (based on the population), considering spectral acceleration at a period of vibration

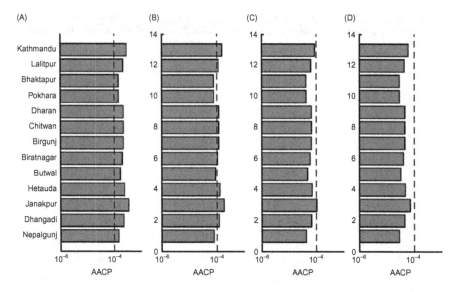

Figure 6.21 AACP for the four NBC models at rock (VS₃₀ = 760 m/s): (A) NBC_IF-1, (B) NBC_IF-2, (C) NBC_IF-3, and (D) NBC_IF-4.

of 0.5 s (thus, in agreement with the intensity measure chosen for the fragility functions presented in Section 6.4). These calculations were performed for rock ($VS_{30} = 760$ m/s) and soil ($VS_{30} = 360$ m/s). Then, using these hazard curves and the fragility functions computed here, the annual collapse probability has been calculated following the procedure proposed by Silva et al. (2014). Figs. 6.21 and 6.22 present the resulting AACP for the 13 cities in Nepal, considering rock and soil, respectively. The vertical dashed line marks the acceptable target risk (i.e., 1.0×10^{-4}).

These results clearly indicate a strong variation of the earthquake risk across Nepal. For example, the AACP in the city of Janakpur is approximately three times higher than in Nepalgunj across the four NBC models. Moreover, it is also possible to conclude that, for areas with a very stiff soil or rock ($VS_{30} > 760$ m/s), there is no need to ensure a WI identical to that of the NBC_IF-4 model. Likewise, regardless of the type of soil, a WI below 1% led consistently to an AACP above the acceptable risk. These findings indicate that the level of safety of structures designed according to the MRT, but with an insufficient WI, may have a safety level below what is internationally recommended. The optimal WI for each city according to the type of

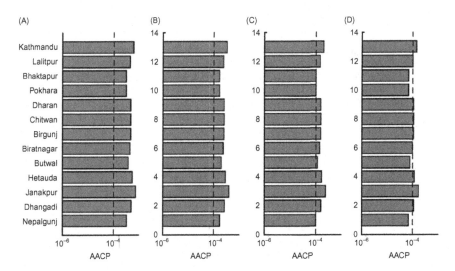

Figure 6.22 AACP for the four NBC models at soil (VS₃₀ = 360 m/s): (A) NBC_IF-1, (B) NBC_IF-2, (C) NBC_IF-3, and (D) NBC_IF-4.

soil is presented in Fig. 6.23. It can be seen from the figure that, for most of the rock sites in Nepal, the minimum required WI is 2.5%, which corresponds to the NBC_IF-3 model considered in this study. The only location which requires a higher WI value of 3.25% is Janakpur. However, soil sites within the Kathmandu valley (e.g., Kathmandu, Lalitpur) also require a higher WI value of 3.25%.

6.6 FINAL REMARKS

RC frame buildings with masonry infills are the most dominant housing type in the urban areas of Nepal. Although reasonably simple code provisions in the form of MRT (NBC 201:1994) have been enforced since 2003, many RC buildings were affected by the 2015 Gorkha earthquake. There are several possible reasons and explanations for earthquake damage in RC buildings at low shaking levels in the Kathmandu valley. An important observation is related to the shear-failure mechanism in the low-rise RC buildings. It appears that the sizes of RC columns and their detailing did not ensure the development of a desirable ductile flexural mechanism. Instead, these buildings performed like wall structures and their seismic response was significantly influenced by the amount of masonry infill walls. Two detailed building surveys with a total of about 140 buildings provided an

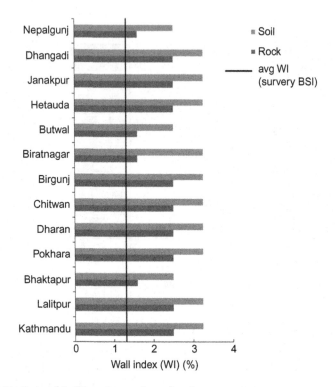

Figure 6.23 Distribution of the WI per city according to the soil type.

opportunity to establish a relationship between the WI and the damage level. These studies have shown that the buildings with a WI value lower than 1.0% experienced significant damage, while the average WI value for surveyed buildings was approximately 1.35% (for three-story buildings). An analytical study involving the development of fragility curves for the RC frame models with four WI values ranging from 0.95% to 3.25% has shown that the buildings with a WI less than 1% have a 40% probability of collapse at spectral accelerations similar to the code-prescribed design earthquake in Kathmandu. The models with WI values of 1.6% and higher showed much smaller collapse probabilities. Finally, a probabilistic assessment of the AACP for the 13 major cities in Nepal has shown that the required values of WI vary considerably across Nepal, but a value of 0.95% leads consistently to risk values above the acceptable threshold.

Based on the various aspects of this study, it can be concluded that the seismic safety of low-rise RC frame buildings with masonry infills in Nepal could be enhanced by prescribing the minimum WI values for low-rise RC buildings addressed by the MRT (NBC 201:1994). The analyses performed in this study are related to three-story buildings. Based on the results presented in Fig. 6.23, the recommended WI values range from 1.6% to 3.25%, depending on the seismic hazard and soil type. Note that the recommended WI value would decrease for two-story buildings. This provision would result in an increase in the amount of walls (in terms of length or thickness) compared to the current construction practice, where an average WI was found to be less than 1.4% (based on the results of building surveys reported in this study).

In terms of the practical application, this provision could be implemented in the MRT (NBC 201:1994) by prescribing wall layout and thickness values for typical building plans that are currently provided in the code. It would be recommended to prescribe thicker walls (e.g., 230 mm for brick masonry) at the building exterior, which would be beneficial both for seismic safety and thermal comfort.

ACKNOWLEDGMENTS

S. Brzev would like to acknowledge contribution of coauthors of the building damage study Bishnu Pandey (BCIT, Canada), Dev Kumar Maharjan (NSET, Nepal), and Carlos Ventura (UBC, Canada). The funding was provided by the 12th CMS Education and Research Fund through the Masonry Institute of British Columbia, Canada. The study was initiated during a postearthquake reconnaissance visit to Nepal, which was sponsored by the Canadian Association for Earthquake Engineering, the British Columbia Institute of Technology, and the UBC Earthquake Engineering Research Facility.

REFERENCES

Basha, S.H., Kaushik, H.B., 2016. Behavior and failure mechanisms of masonry-infilled RC frames (in low-rise buildings) subject to lateral loading. Eng. Struct. 111, 233–245.

Brzev, S., Scawthorn, C., Charleson, A.W., Allen, L., Greene, M., Jaiswal, K., et al., 2013. GEM Building Taxonomy Version 2.0. GEM Technical Report 2013-02 V1.0.0. GEM Foundation, Pavia.

Brzev, S., Pandey, B., Maharjan, D.K., Ventura, C. 2017. Seismic vulnerability index for low rise composite reinforced concrete and masonry buildings in Nepal. In: Proceedings of 16th World Conference on Earthquake Engineering, Santiago, Chile.

Brzev, S., Pandey, B., Pao, J., 2017. Urban housing: performance of reinforced concrete buildings. Report on Earthquake Reconnaissance of the M7.8 Gorkha, Nepal Earthquake on April 25, 2015 and Its Aftershocks. Canadian Association of Earthquake Engineering, Vancouver.

Calvi, G.M., 1999. A displacement-based approach for vulnerability evaluation of classes of buildings. J. Earthquake Eng. 3 (3), 411–438.

CBS, 2012. National Report, National Planning Commission Secretariat. Central Bureau of Statistics, Kathmandu.

Chaulagain, H., Rodrigues, J., Jora, J., Spacone, E., Varum, H., 2013. Seismic response of current RC buildings in Nepal: a comparative analysis of different design/construction. Eng. Struct. 49, 281–294.

Chaulagain, H., Rodrigues, H., Silva, V., Spacone, E., Varum, H., 2015. Seismic risk assessment in Nepal. Nat. Hazard 78 (1), 583–602.

Chitrakar, G.R., Pandey, M.R., 1986. Historical Earthquakes of Nepal. Bull. Geol. Soc. Nepal (4), 7–8.

Crisafulli, F. 1997. Seismic Behaviour of Reinforced Concrete Structures with Masonry Infills (Ph.D. dissertation). University of Canterbury, New Zealand.

Crowley, H., Pinho, R., Bommer, J.J., 2004. A probabilistic displacement-based vulnerability assessment procedure for earthquake loss estimation. Bull. Earthq. Eng. 2, 173–219.

Dixit, A., 2004. Promoting safer building construction in Nepal. In: Proceedings of the 13th World Conference on Earthquake Engineering, Vancouver, Canada.

EERI, 2016. M7.8 Gorkha, Nepal Earthquake on April 25, 2015 and Its Aftershocks. Earthquake Engineering Research Institute, Oakland, CA, <http://www.eeri.org/>.

Elnashai, A.S., Elghazouli, A.Y., 1993. Performance of composite steel/concrete members under earthquake loading. Part I: analytical model. Earthquake Eng. Struct. Dyn. 22, 315–345.

Gautam, D., Rodrigues, H., Bhetwal, K.K., Neupane, P., Sanada, Y., 2016. Common structural and construction deficiencies of Nepalese buildings. Innovat. Infrastruct. Solut. 1 (1), 1–18.

Geotechnical Extreme Event Reconnaissance Association (GEER), 2015. Gorkha (Nepal) earthquake of April 25, 2015 and related shaking sequence. GEER Association Report No. GEER-040, Version 1.1. <www.geerassociation.org/component/geer_reports/?view=geerreports&id=26&layout=default>.

Grünthal, G., Musson, R.M.W., Schwarz, J., Stucchi, M. (Eds.), 1998. European Macroseismic Scale 1998, vol. 15. Cahiers du Centre Européen de Géodynamique et de Séismologie, Luxembourg.

Gulkan, P., Sozen, M.A., 1999. Procedure for determining seismic vulnerability of building structures. ACI Struct. J. 96 (3), 336–342.

Guragain, R., Shrestha, S.N., Maharjan, D.K., Pradhan, S. 2017. Building damage patterns of non-engineered masonry and reinforced concrete buildings during April 25, 2015 Gorkha Earthquake in Nepal. In: Proceedings of 16th World Conference on Earthquake Engineering, Santiago, Chile.

Hassan, A.F., Sozen, M.A., 1997. Seismic vulnerability of low-rise buildings in region with infrequent earthquakes. ACI Struct. J. 94 (1), 31–39.

Jalayer, F., Cornell, C., 2002. Alternative nonlinear demand estimation methods for probability-based seismic assessments. Earthquake Eng. Struct. Dyn. 38 (8), 951–972.

Lava, J., Avouac, J.P., 2000. Active folding of fluvial terraces across the Siwaliks Hills, Himalayas of Central Nepal. J. Geophys. Res. 105, 5735–5770.

Mander, J.B., Priestley, M.J.N., Park, R., 1988. Theoretical stress-strain model for confined concrete. J. Struct. Eng. ASCE 114 (8), 1804–1826.

Martinez-Rueda, J.E., 1997. Energy Dissipation Devices for Seismic Upgrading of RC Structures (Dissertation). Imperial College, University of London, London, UK.

Martín Tempestti, J., Stavridis, A., 2017. Simplified analytical method to predict the failure pattern of infilled RC frames. In: Proceedings of 16th World Conference on Earthquake Engineering, Santiago, Chile.

Mehrabi, A.B., P.B. Shing, M.P. Schuller, Noland, J.L., 1994. Performance of masonry-infilled R/C frames under in-plane lateral loads. Report No. CU/SR-94/6, University of Colorado at Boulder, USA.

Meli, R., Brzev, S., et al., 2011. Seismic Design Guide for Low-Rise Confined Masonry Buildings. Earthquake Engineering Research Institute, Oakland, CA.

Menegotto, M., Pinto, P.E., 1973. Method of analysis for cyclically loaded RC plane frames including changes in geometry and non-elastic behaviour of elements under combined normal force and bending. Symposium on the Resistance and Ultimate Deformability of Structures Acted on by Well Defined Repeated Loads. International Association for Bridge and Structural Engineering, Zurich.

Moroni, M.O., Astroza, M., Caballero, R., 2000. Wall density and seismic performance of confined masonry buildings. TMS J. 18 (1), 79–86.

Moroni, M.O., Astroza, M., Acevedo, C., 2004. Performance and seismic vulnerability of masonry housing type in Chile. J. Perform. Construct. Facilities ASCE 18 (3), 173–179.

Murty, C.V.R., Brzev, S., Faison, H., Comartin, C.D., Irfanoglu, A., 2006. At risk: the seismic performance of reinforced concrete frame buildings with masonry infills. World Housing Encyclopedia. Earthquake Engineering Research Institute, Oakland, CA. Available as Publication WHE-2006-03. <www.world-housing.net/tutorials/reinforced-concrete-tutorials>.

NBC 105, 1994. Seismic design of buildings in Nepal. Nepal National Building Code. Department of Urban Development and Building Construction, Kathmandu.

NBC 201, 1994. Mandatory rules of thumb—reinforced concrete buildings with masonry infill. Nepal National Building Code. Department of Urban Development and Building Construction, Kathmandu.

NBC 203, 1994. Guidelines for Earthquake Resistant Building Construction: Low Strength Masonry. Department of Urban Development and Building Construction, Kathmandu.

NPC, 2015. Nepal Earthquake 2015: Post Disaster Needs Assessment, Executive Summary. National Planning Commission Government of Nepal, Kathmandu. Available at <http://www.npc.gov.np/>.

Pagani, M., Monelli, D., Weatherill, G., Danciu, L., Crowley, H., Silva, V., et al., 2014. OpenQuake engine: an open hazard (and risk) software for the Global Earthquake Model. Seismol. Res. Lett. 85 (3), 692–702.

Pradhan, P.L., 2009. Composite Action of Brick Infill Wall in RC Frame Under In-plane Lateral Load (Dissertation). Institute of Engineering, Pulchowk Campus, Tribhuvan University, Nepal.

Priestley, M.J.N., 1997. Displacement-based seismic assessment of reinforced concrete buildings. J. Earthquake Eng. 1 (1), 157–192.

Rana, B.J.B., 1935. Nepal ko Maha Bhukampa [Great Earthquake of Nepal]. Jorganesh Press, Kathmandu.

SeismoSoft., 2015. SeismoStruct—a computer program for static and dynamic nonlinear analysis of framed structures, (online). <http://www.seismosoft.com/>.

Shah, P., Pujol, S., Kreger, M., Irfanoglu, A., 2017. 2015 Nepal Earthquake, concrete international. Am. Concrete Inst. 39 (3), 42–49.

Shrestha, B., Dixit, A.M., 2008. Standard design for earthquake resistant buildings and aid to building code implementation in Nepal. In: Proceedings of the 14th World Conference on Earthquake Engineering, Beijing, China.

Silva, V., Crowley, H., Pagani, M., Monelli, D., Pinho, R., 2014. Development of the OpenQuake engine, the global earthquake model's open-source software for seismic risk assessment. Nat. Hazards 72 (3), 1409–1427.

Smyrou, E., Blandon, C., Antoniou, S., Pinho, R., Crisafulli, F., 2011. Implementation and verification of a masonry panel model for nonlinear dynamic analysis of infilled RC frames. Bull. Earthq. Eng. 9 (5), 1519–1534, <http://dx.doi.org/10.1007/s10518-011-9262-6>.

Spence, R., 2004. Earthquake protection: the need for legislation to strengthen high-risk buildings. In: Proceedings of 13th World Conference on Earth Engineering, Vancouver, Canada.

Thapa, N., 1988. Bhadau Panch Ko Bhukampa [in Nepali]. Central Disaster Relief Committee, Kathmandu.

Thapa, D.R., Guoxin, W., 2013. Probabilistic seismic hazard analysis in Nepal. Earthq. Eng. Eng. Vib. 12, 577–586.

UNCRD, 2008. Handbook on Building Code Implementation. United Nations Centre for Regional Development, Disaster Management Planning Hyogo Office, Hyogo.

UNDP, 2010. Recommendations for construction of earthquake safer buildings—earthquake risk reduction and recovery preparedness programme for Nepal. UNDP/ERRRP-Project: NFP/07/010, Department of Urban Development and Building Construction, Babarmahal, Kathmandu.

USGS, 2015. Gorkha Earthquake M7.8—36 km E of Khudi, Nepal: General Summary web page, United States Geological Survey. <https://earthquake.usgs.gov/earthquakes/eventpage/us20002926>.

FURTHER READING

Chamlagain, D., Hayashi, D., 2004. Numerical simulation of fault development along NE–SW Himalayan profile in Nepal. J. Nepal Geol. Soc. 29, 1–11.

Chamlagain, D., Hayashi, D., 2007. Neotectonic fault analysis by 2D finite-element modeling for studying the Himalayan fold-and-thrust belt in Nepal. J. Asian Earth Sci. 29, 473–489.

Chaulagain, H., Rodrigues, H., Silva, V., Spacone, E., Varum, H., 2016. Earthquake loss estimation for the Kathmandu Valley. Bull. Earthquake Eng. 14 (1), 59–88.

Dahal, R.K., 2006. Geology for Technical Students. Bhrikuti Academic Publications, Kathmandu, p. 756.

Karmacharya, U., 2017. Assessment of Wall Density Was Seismic Vulnerability Indicator for RC Buildings in Nepal (Dissertation). Understanding and Managing Extremes School, Pavia, Italy.

Khattri, K.N., Wyss, M., 1978. Precursory variation of seismicity rate in the Assam areas, India. Geology 6, 685–688.

Silva, V., Crowley, H., Bazzurro, P., 2016. Exploring risk-targeted hazard maps for Europe. Earthquake Spectra 32 (2), 1165–1186.

CHAPTER 7

Past and Future of Earthquake Risk Reduction Policies and Intervention in Nepal

Dipendra Gautam
University of Molise, Campobasso, Italy

7.1 INTRODUCTION

Earthquake risk reduction needs integrated efforts from the stakeholders and vitally needs a well-defined framework in terms of an act or policy specifies the duties and responsibilities and identifies the individual, community, and government efforts or the framework to collaborate. Disaster prone countries need exhaustively defined frameworks to minimize the casualties during and after the disasters. Moreover, a road map developed in terms of policies often helps minimize ambiguities between various stakeholders involved in the process. Ambiguous policies lead to aggravated disaster effects, as in the case of 2015 Gorkha earthquake. Immediately after the Gorkha earthquake, the government decided to follow a one-door policy in response, relief, and recovery, but due to continuous pressure and criticisms from nongovernment sectors, this policy was later suspended. This suspension allowed agencies to directly act in situ, which ultimately led to concentrating some 10,000 agencies in a single district. However, their performance was questionable, as the victims seldom got relief support or temporary shelters from the agencies. Immediately after the earthquake, many countries from across the world decided to send relief support as well rapid response forces. The unrestricted access to the damage locations and uncoordinated responses also created problems like concentration of relief support in some accessible areas and interruption in the rescue process. Due to lack of a proper media management strategy, rumors created panic among the victims for several days. Apart from this, some cases of robbery and sex trafficking were also reported immediately after the Gorkha earthquake by several online and print media. All the consequences were due to either lack or failure in implementation of policies. The government declared 14 out

Impacts and Insights of the Gorkha Earthquake. DOI: http://dx.doi.org/10.1016/B978-0-12-812808-4.00007-9

of 31 affected districts as crisis-hit areas, however, unrestricted access and small-scale distribution of relief support were not checked.

Nepal progressed considerably in recent decades in terms of risk reduction, more precisely, in earthquake risk reduction, and some changes in terms of construction of resilient buildings that could prevent loss of life can be observed in some urban areas. However, the most of the population still remains in the rural areas, where vulnerable buildings are dominant and awareness is seldom reached. In addition to this, drill campaigns like the Drop-Cover-Hold On have reached certain levels of the population, although such drills may not have a commendable impact. Two school children in the neighboring village of Kathmandu lost their lives when they went from outdoors into their home from to hide under a table. Awareness campaigns, if not localized and do not consider the vulnerability of structures, can give rise to more detrimental consequences as highlighted by such evidence. Considering the historical evidence of damage, casualties, and success and failure of policies, this chapter critically reviews the policies related to earthquake risk reduction and presents some insights for future improvements.

7.2 EARTHQUAKE RISK REDUCTION POLICIES AND INTERVENTIONS IN NEPAL: OVERVIEW AND CRITIQUES

7.2.1 Natural Calamity Relief Act (1982)

The Natural Calamity Relief Act (NCRA) was promulgated by the then-majesty's government in 1982 to operate relief work and protect life and property. The NCRA was the first disaster risk management act in the entire South Asian region; however, it remains static, even after three and half decades since its promulgation. The NCRA has provisioned declaration of a disaster area, various orders the government can give, control of entry by foreigners, formation of Central Natural Disaster Relief Committee (CNDRC) under the chairmanship of the home minister, and formation of several other subcommittees, like a relief and treatment subcommittee and a supplies, shelter, and rehabilitation subcommittee. In addition to this, the NCRA endorses formation of regional-, district-, and local-level disaster relief committees. In every relief committee level, the aid fund is provided by the NCRA. All the funds are expected to be circulated under "one-door policy," so that the government could effectively allocate the funds per its priority.

The NCRA is a basic document, so that most of the aspects in terms of community-based approaches, cooperative approaches, countermeasures for impending disasters, identification of potential hazards, among others are lacking in this act. In fact, the NCRA is the document for postdisaster priorities only and predisaster initiatives as well as during-the-event interventions cannot be found. It does not specify the specific interventions for earthquake disaster either, although earthquake was already known as the most fatal disaster in Nepal during its promulgation.

7.2.2 National Building Act (1988)

To regulate the building construction and protect buildings from earthquakes, fire, and other disasters, Nepal formulated the National Building Act of 1998 (http://moud.gov.np/wp-content/uploads/2016/08/building-act-2055-1998-english.pdf). The National Building Act (1998) proposes a committee to prepare building codes for four different types of buildings: modern reinforced concrete constructions, buildings with plinth area more than 1000 ft² and structural span of more than 4.5 m, buildings with plinth area up to 1000 ft² and structural span up to 4.5 m, and houses other than those just mentioned. This act assures that all buildings should be compliant with the building code. Although the act assures that the buildings will be compliant to the building codes, implementation is largely lacking in Nepal. Until 2014, only 20 out of 58 municipal centers had endorsed the buildings code in terms of mandatory rules of thumb. Construction in the rest of the 3913 villages is still beyond the guidelines, leading to majority of existing building stock in Nepal highly vulnerable to disasters. The current building code is largely confined to earthquake and fire, whereas Nepal is prone to multiple hazards. The current building code was formulated in 1994 after the 1988 earthquake and implementation was started only in 2003 in some municipal centers. Due to a high dependence on the Indian Standard Codes, most of the current designs and constructions in Nepal are based on Indian Standard Codes.

7.2.3 National Urban Policy (2007)

The National Urban Policy (2007) was formulated to address the problems related to the haphazard urban development practices in Nepal. This policy highlighted the planned urban development with the help of master plans; however, even now, it is seldom practiced. The focus is on economic development and formulation of north-south and east-west access corridors; meanwhile, risk reduction is not thoroughly addressed by National Urban Policy (2007).

7.2.4 National Strategy on Disaster Risk Management (2009)

The National Strategy on Disaster Risk Management has set five high-priority areas including disaster risk management at the national and local levels; improvement of potential risk assessment; identification, monitoring, and preparedness; use of knowledge; and new ideas for a culture of resilience and safety, minimization of risk factors, and effective disaster preparedness for effective response. Although high-priority areas are well identified and defined, implementation and endorsement of new ideas is largely lacking in Nepal. For instance, the seismic hazard mapping and identification of potential risk is based on the formulation of the Building Code Development Project (BCDP, 1994) and most of the recent studies (e.g., Thapa and Guoxin, 2013; Chaulagain et al., 2015) are not introduced until late. In terms of loss estimation, JICA (2002) and Chaulagain et al. (2016) updated the risk scenario at least for Kathmandu valley; however, such studies are not introduced or endorsed while formulating new policies or planning the development. Apart from earthquakes, multihazard risk assessment is lacking in Nepal, and the vulnerability of the existing building stock is not well identified; this scenario is leading to enormous losses in every notable disaster.

7.2.5 National Shelter Policy (2012)

In 2012 the National Shelter Policy of 1996 was amended to address the social and political changes in Nepal. The National Shelter Policy ensured housing for the deprived, homeless and, displaced families from the areas of major project construction sites as well as natural disasters. In addition to this, provision of alternative settlements for the families residing in unsafe areas is also assured by the National Shelter Policy to downscale the risks and damages due to impending natural and anthropogenic disasters. Although the shelter policy assures public safety at the policy level, implementation has not become common in Nepal. Except for the project construction sites, people are not relocated considering the disasters. For example, people were still residing in the same locations after the 2014 landslide in Jure area of Sindhupalchowk district, and the same area was severely affected by the 2015 Gorkha seismic sequence.

7.2.6 Urban Planning and Development Act (2015)

The Urban Planning and Development Act (2015) was formulated to address the new challenges posed by the Gorkha earthquake of April 25, 2015. This act proposes technical personnel in metropolitan cities, submetro areas, municipalities, and other urban centers for drawing approval and construction supervision. Structural design along with the architectural plan is made compulsory by this act; however, such practice is limited only to reinforced concrete structures in urban area Per this act, construction approval is given in two tiers: the first up to the Damp Proofing Course (DPC) and verification is done after this and a decision is taken for next phase construction. One important feature of this act is the provision of open spaces on a local scale. During Gorkha seismic sequence, the Kathmandu valley population hardly got a chance to establish their tents and tarps in open spaces. Noting this scenario, the Urban Planning and Development Act (2015) provisioned conservation of open spaces on a local scale. Rural settlements consist sparsely distributed buildings in Nepal; on the contrary, urban centers are crowded and no open space is available in most of such urban areas in the case of an emergency. Land use planning is one of the biggest challenges in urban areas of Nepal, as most of the settlements established in recent decades are on arable lands used for agriculture for centuries. Without engineering judgment, such arable lands are used for building construction sites and the damage accumulation is unavoidable, as in the case of 2015 Gorkha earthquake (see Gautam et al., 2016; Gautam and Chaulagain, 2016).

7.3 THE WAY FORWARD

Some other existing acts, regulations, policies, and bylaws are indirectly related to earthquake risk reduction. For example, the Lands Act (1964) has ensured the secure settlements and purposeful land use planning only. In some areas of Nepal, risk-sensitive land use planning is being practiced; however, such risk reduction strategies have not gained momentum at the national level. The present need in Nepal is multihazard mapping and relocation of hazard-prone communities. Identification of risk in Nepal is largely confined to some urban centers and limited regions, and most of the country's rural areas are not considered for hazard identification and risk reduction strategies. Nearly 37% of the total area in Nepal is urbanized and most of such urban centers are promoted from the villages directly without any infrastructural upgradation as well as planning. Until 2014, only 17% urban areas existed in Nepal and, after endorsing federalism, another 20% of areas were declared as urban areas in 2015 and 2016. Some crucial ways forward in terms of earthquake risk reduction can be postulated as follows:

1. Nepal largely depends on the Indian building code or the marginal Nepal Building Code, which was formulated during 1994. Most of the provisions in these buildings codes do not address the local issues arising at the community level. For example, the northwestern part of the Kathmandu valley sustained severe damage during the Gorkha earthquake as the settlement was in previous river course. The Nepal Building Code needs to provide site-specific design guidelines, so a new building code is must for Nepal.

2. Apart from earthquakes, interacting hazards are frequent and widely noticed in Nepal. The challenge is to identify possible interacting hazards, like earthquakes followed by floods, avalanches, landslides, fire, among others. In doing so, a detailed assessment and multidisciplinary approach is required. Resilience against only earthquakes does not address the current multihazard risk in Nepal.

3. Current understating of seismic hazard is not exhaustive, both historical and instrumental seismicity should be rigorously improved to demarcate the precise hazard level. Limited database and constraints have minimized the precision level of seismic hazard assessments. In addition to this, far-field events in the Hindu-Kush-Himalaya may affect in Nepal, so improvement in terms of seismic hazard assessment is needed. For example, the 2001 Bhuj earthquake was also felt in Nepal ($>$1500 km).

4. Many Nepali settlements are situated in fertile, loose organic soil deposits due to economic and agricultural values in the past. For example, Kathmandu itself is an alluvial basin that suffers severely even in case of far-field earthquakes (e.g., 1833, 1934, 1988, and 2015). This ultimately highlights the necessity of region-specific seismic considerations. Microzonation schemes are urgently required for the major settlements, whereas identification of multihazard risk and relocation of sparsely located households in the middle and high mountains should be prioritized.

5. The central seismic gap is believed to dormant for more than 500 years, and no major earthquakes have occurred in country's western part. Most of the areas in the western middle and high mountains consist of highly vulnerable rubble stone buildings; hence, the

losses in the case of major earthquakes in western Nepal are several times more severe than what occurred during 1934 and 2015 earthquakes.

6. An important aspect in earthquake risk reduction is the lack of implementation of related acts and regulations. Even the marginal buildings code is not made mandatory to all the settlements. This led to continued construction of highly vulnerable structures, so ~70% or more of the buildings in Nepal are highly vulnerable. A framework for either reconstruction or strengthening is required urgently. However, strengthening local buildings is challenging due to material and technology constraints and the economic status of owners, so the government of Nepal needs to start feasible and resilient housing construction programs in the areas not affected by the Gorkha earthquake, too.

7. Nearly 70% of engineered residential buildings in Nepal do not fulfill the design expectations due to compromised material quality and workmanship. Therefore, urban engineered buildings also need to be strengthened. To facilitate strengthening, a code provision is needed in Nepal.

8. Capacity building is crucial aspect in resilient constructions as well as policy making. Currently there is no sufficient manpower in structural earthquake engineering trained by universities in Nepal. A practice-oriented and real-time problem-solving education system for engineers as well as policy makers can bring positive changes.

9. Identification, justification, and promotion of indigenous knowledge related to disasters is not considered on a broad scale. For example, traditional constructions in Nepal in various regions have some seismic features; however, such technologies are rarely well documented. Other indigenous practices, like the trust system practiced in several communities, are important in postearthquake relief, rehabilitation, and reconstruction. The current urbanization trend has threatened the low-cost traditional values and technologies. A framework is needed to preserve indigenous knowledge in both urban and rural areas.

10. Historically, Nepali settlements included open spaces in every community; however, due to commercialized urbanization and haphazard growth, most of the open spaces no longer exist. Construction and maintenance of large water bodies in

Figure 7.1 Location of open spaces in Kathmandu valley. As shown in the figure, most of the densely populated areas in the valley do not have sufficient public space in the case of emergency.

communities for firefighting also no longer exist, leading to delayed response in case of postearthquake fire outbreak. Now, only 83 identified open spaces (see Fig. 7.1) remain in the Kathmandu valley for nearly 4 million inhabitants. Lack of open space was widely noticed during the 2015 Gorkha earthquake, so most of the people fixed their tents and tarps on the streets, which led to delayed emergency response.

11. Government-level failure in terms of media management and state of implementation of rules and regulations increased the trauma among the ordinary people during the Gorkha earthquake, as rumors of strong earthquakes along with their timing spread widely through social media, which created complete chaos for some weeks after the main shock of April 25, 2015. Nepal did not declare a state of emergency, so unrestricted access led to exaggerated reports, causing mental distress among the victims. At the same time, funds were largely mismanaged, as relief distribution was not properly checked. This led to multiple relief supplies to accessible communities and none to the inaccessible ones. A one-door policy for relief distribution, rehabilitation, and reconstruction should be enforced strictly to avoid such disparities.

12. Nepal lacks a postdisaster reconstruction framework and regulations, this led in delayed reconstruction efforts after the 2015 Gorkha

earthquake. Even the victims of the 2011 Sikkim-Nepal earthquake did not receive relief support until 2016; surely, the victims of the 2015 Gorkha earthquake would face a similar situation. Regarding reconstruction, the government depended too much on international funding rather than finding local solutions, which led to further delays. Social institutions like trust system and labor exchange are widely practiced in Nepal, and cooperative movements have reached to rural levels, too. So, local solutions were possible for mass reconstruction in Nepal. The government lured the victims with subsidies; however, the process was lengthy; the government also lacked a sure of reconstruction model; and the victims were obliged to stay under the tarps for two years. A reconstruction plan for all possible disasters in Nepal is needed. and strict implementation is required for effective earthquake risk reduction strategies.

7.4 FINAL REMARKS

The sum of the overview and critiques highlights that formulation of policies alone are insufficient for earthquake risk reduction. Moreover, implementation of existing policies and guidelines results in some changes, as expected, even if the policies and guidelines are in a primitive stage. Together with the formulation of policies and regulations, Nepal needs to focus on implementation. As integrated risk reduction approaches are not observed, formulation of integrated risk reduction policies and regulations as an act is necessary for Nepal. In doing so, a multihazard risk assessment at the local scale is important as highlighted by the localized impacts of several disasters in the past. Apart from this, Nepal needs a new building code, compliant with the impending seismic hazard as well as other natural and anthropogenic hazards. Due to unrestrained development, future earthquakes would be more detrimental than the past ones due to increment in exposure and inherent vulnerability of buildings and other infrastructures; therefore, immediate policy changes and strict implementation are need to assure the safety of the public in Nepal.

ACKNOWLEDGMENTS

The author is grateful to Professor Hugo Rodrigues for his efforts in reviewing this chapter. Any opinions or conclusions expressed in this chapter are those of author and do not necessarily reflect the opinion of any affiliated institution.

REFERENCES

Building Code Development Project (BCDP), 1994. Seismic Hazard Mapping and Risk Assessment for Nepal. UNDP/UNCHS (Habitat) Subproject: NEP/88/054/21.03. Ministry of Housing and Physical, Planning, Government of Nepal, Kathmandu.

Chaulagain, H., Rodrigues, H., Silva, V., Spacone, E., Varum, H., 2015. Seismic risk assessment and hazard mapping in Nepal. Nat. Hazards 78 (1), 583–602.

Chaulagain, H., Rodrigues, H., Silva, V., Spacone, E., Varum, H., 2016. Earthquake loss estimation for the Kathmandu valley. Bull. Earthquake Eng., 14 (1), 59–88.

Gautam, D., Chaulagain, H., 2016. Structural performance and associated lessons to be learned from world earthquakes in Nepal after 25 April 2015 (M_W 7.8) Gorkha earthquake. Eng. Failure Anal., 68, 222–243.

Gautam, D., Rodrigues, H., Bhetwal, K.K., Neupane, P., Sanada, Y., 2016. Common structural and construction deficiencies in Nepalese buildings. Innovat. Infrastruct. Solut., 1, 1. Available from: http://dx.doi.org/10.1007/s41062-016-0001-3.

Japan International Co-operation Agency (JICA). 2002. The Study on Earthquake Disaster Mitigation in the Kathmandu Valley, Kingdom of Nepal. Japan International Cooperation Agency and the Ministry of Home Affairs of Nepal, Kathmandu, vol. I–IV.

Thapa, D.R., Guoxin, W., 2013. Probabilistic seismic hazard analysis in Nepal. Earthquake Eng. Eng. Vibrat., 12, 577–586.

Printed in the United States
By Bookmasters